Industrial Applications of High-Performance Computing

Best Global Practices

Chapman & Hall/CRC Computational Science Series

SERIES EDITOR

Horst Simon
Deputy Director
Lawrence Berkeley National Laboratory
Berkeley, California, U.S.A.

PUBLISHED TITLES

COMBINATORIAL SCIENTIFIC COMPUTING
Edited by Uwe Naumann and Olaf Schenk

CONTEMPORARY HIGH PERFORMANCE COMPUTING: FROM PETASCALE TOWARD EXASCALE
Edited by Jeffrey S. Vetter

CONTEMPORARY HIGH PERFORMANCE COMPUTING: FROM PETASCALE TOWARD EXASCALE, VOLUME TWO
Edited by Jeffrey S. Vetter

DATA-INTENSIVE SCIENCE
Edited by Terence Critchlow and Kerstin Kleese van Dam

THE END OF ERROR: UNUM COMPUTING
John L. Gustafson

FUNDAMENTALS OF MULTICORE SOFTWARE DEVELOPMENT
Edited by Victor Pankratius, Ali-Reza Adl-Tabatabai, and Walter Tichy

THE GREEN COMPUTING BOOK: TACKLING ENERGY EFFICIENCY AT LARGE SCALE
Edited by Wu-chun Feng

GRID COMPUTING: TECHNIQUES AND APPLICATIONS
Barry Wilkinson

HIGH PERFORMANCE COMPUTING: PROGRAMMING AND APPLICATIONS
John Levesque with Gene Wagenbreth

HIGH PERFORMANCE PARALLEL I/O
Prabhat and Quincey Koziol

HIGH PERFORMANCE VISUALIZATION:
ENABLING EXTREME-SCALE SCIENTIFIC INSIGHT
Edited by E. Wes Bethel, Hank Childs, and Charles Hansen

INDUSTRIAL APPLICATIONS OF HIGH-PERFORMANCE COMPUTING:
BEST GLOBAL PRACTICES
Edited by Anwar Osseyran and Merle Giles

PUBLISHED TITLES CONTINUED

INTRODUCTION TO COMPUTATIONAL MODELING USING C AND OPEN-SOURCE TOOLS
José M Garrido

INTRODUCTION TO CONCURRENCY IN PROGRAMMING LANGUAGES
Matthew J. Sottile, Timothy G. Mattson, and Craig E Rasmussen

INTRODUCTION TO ELEMENTARY COMPUTATIONAL MODELING: ESSENTIAL CONCEPTS, PRINCIPLES, AND PROBLEM SOLVING
José M. Garrido

INTRODUCTION TO HIGH PERFORMANCE COMPUTING FOR SCIENTISTS AND ENGINEERS
Georg Hager and Gerhard Wellein

INTRODUCTION TO REVERSIBLE COMPUTING
Kalyan S. Perumalla

INTRODUCTION TO SCHEDULING
Yves Robert and Frédéric Vivien

INTRODUCTION TO THE SIMULATION OF DYNAMICS USING SIMULINK®
Michael A. Gray

PEER-TO-PEER COMPUTING: APPLICATIONS, ARCHITECTURE, PROTOCOLS, AND CHALLENGES
Yu-Kwong Ricky Kwok

PERFORMANCE TUNING OF SCIENTIFIC APPLICATIONS
Edited by David Bailey, Robert Lucas, and Samuel Williams

PETASCALE COMPUTING: ALGORITHMS AND APPLICATIONS
Edited by David A. Bader

PROCESS ALGEBRA FOR PARALLEL AND DISTRIBUTED PROCESSING
Edited by Michael Alexander and William Gardner

SCIENTIFIC DATA MANAGEMENT: CHALLENGES, TECHNOLOGY, AND DEPLOYMENT
Edited by Arie Shoshani and Doron Rotem

Industrial Applications of High-Performance Computing

Best Global Practices

Edited by

Anwar Osseyran
SURFsara BV
Netherlands

Merle Giles
NCSA, University of Illinois
Urbana, Illinois, USA

CRC Press is an imprint of the
Taylor & Francis Group, an informa business
A CHAPMAN & HALL BOOK

CRC Press
Taylor & Francis Group
6000 Broken Sound Parkway NW, Suite 300
Boca Raton, FL 33487-2742

Printed on acid-free paper
Version Date: 20150227

International Standard Book Number-13: 978-1-4665-9680-1 (Hardback)

Visit the Taylor & Francis Web site at
http://www.taylorandfrancis.com

and the CRC Press Web site at
http://www.crcpress.com

Contents

Part II Case Studies

Part III Industry Developments and Trends

Foreword

YOU ARE ABOUT TO ENTER THE EMERGING WORLD OF THE SYNTHESIS of mankind and machine. This is not of the type usually portrayed in science fiction with horrific results. Rather it is of the constructive synergies that are redefining civilization through the combination of human creativity and the power of supercomputing. For tens of thousands of years, empiricism, essentially trial and error, led us through the series of original innovations. Only in the last 3000 years did the slow quest to acquire knowledge through science and combine it with engineering allow humanity to begin to master its environment through understanding and technology. But in a single lifetime a third pillar of learning and invention has been added to theory and experiment—digital modeling of complex phenomenology due to the advent of the supercomputer. With the supercomputer, we can understand the past, control the present, and, in certain restrictive but important cases, predict the future. The supercomputer allows us to analyze massive data from observations and simulate possible future engineering feats through modeling in the abstraction of mathematics. The litany of accomplishments through the use of supercomputers on which we rely so heavily today is vast to the point of being mind-numbing. Yet, only through this powerful new medium of exploration are we able to address such universal and fundamental questions as climate change, controlled fusion and renewable energy, cosmology and exoplanets, microbiology and drug design, automobile and aircraft development, new materials like graphene and carbon nanotubes, and semiconductor electronics including, perhaps ironically, those from which supercomputers themselves are fabricated. Two fundamental discoveries that help us understand the universe in which we live are that of the Higgs Boson particle that gives mass to matter and Dark Energy that is accelerating the expansion of the universe rather than allowing it to slow down. Truly how we see our reality and mold it to our needs is guided to a great degree by high-performance computing and related infrastructure we build, and the way we apply them to the many challenges faced by humanity.

The success of this new fulcrum of human endeavor, this crucible of mankind's creativity, is dependent on the triad of partnership among computing technology, modeling and analytics, and scientific knowledge acquisition manifested as engineering innovation. Although early concepts for mechanized computing were devised by such geniuses as Pascal, Leibnitz, and Babbage, it was not until the late nineteenth century that perhaps the first supercomputer, the Hollerith machine, created to perform the 1890 U.S. census, first demonstrated practical, indeed essential, application. By the decade of the 1930s, electromechanical and then vacuum tube-based analog computers combined the three essential elements into this partnership to solve physical systems described as sets of first-order differential equations. The revolution of digital electronics from initial relays (Harvard Mk 1), to vacuum tubes (ENIAC and Whirlwind), to transistors (IBM 7090), and to early integrated circuit technology (CRAY 1) finally concluding with multi-core and GPU ensembles spanned a performance gain of 12 orders of magnitude in one lifetime, the single most dramatic revolution of any in the history of human technology.

The book you are holding in your hands advances this story from the realm of this oft-told precession of technology steps to the real-world impact of the supercomputer on the diversity of science, engineering, industrial and commercial domains as well as the breadth of the planet we inhabit. It extends our preconceived but flawed idea of supercomputers as the limited purview of a few rich nations to an almost universally accessible medium of economic and cultural progress. Supercomputers are now the principal tool of science accomplishment whether spanning the cosmos or focusing on the subatomic. Supercomputers are now the principal tools of engineering design from kilometer-high skyscrapers and kilometer-long bridges to all forms of transportation to new materials and medical treatments and far more. Supercomputers are now the principal means of enhancing information gathering, archiving, structuring, searching, retrieval, and presentation. Supercomputers are now enabling the fastest growing areas of entertainment for games and movies through visual effects. Supercomputers are the principal means of gathering new knowledge through data mining and machine learning.

The larger lesson of this book is that supercomputers occupy the pursuits of many of the nations of the earth, large or small, creating an equality of opportunity for all who seek to benefit from its power through creative application. As will be seen through these pages, any nation with modest resources can gain most of the benefits of this tool. True, only one nation at any one time will have the fastest computer in the world (today it is the Chinese Tianha-2). But as is apparent from the TOP500 List of the world's fastest computers as measured by the so-called High-Performance Linpack (HPL) benchmark, more than 80% of all such machines are within a factor of 2× to 4× of each other; all almost essentially the same. This means that any nation can eventually join the club, acquire the capability afforded by supercomputing for its industry and economy, and devise the expertise to exploit it to their benefit as are many such nations today.

Interestingly, a second aspect of the lesson presented so vividly in this book is the counter-intuitive idea that competitiveness can be enhanced through international cooperation. The not-invented-here syndrome can be fatal to excelling in the exploitation of supercomputing. Only through cooperation across the nations have such successes as Linux, MPI, and a host of environment tools permitted a degree of standardization supporting individual (national) programs of computational progress. Even as it is argued that supercomputers can advance competitiveness, it is through the sharing of common means, methods, and tools that all can benefit. If every nation had to reinvent their own software stack or programming models, no one would accomplish almost anything of real value. Thus, competition is built on top of cooperation, unique accomplishments leverage shared common environments. Even the supercomputers themselves, through the advances in commodity cluster computers, have made it possible for almost any institution to deploy its own supercomputer through in-house integration of commodity subsystems acquired from a multiplicity of sources. And because of the effects of economy of scale and the mass marketing of those same subsystems, overall system deployment costs can be remarkably low. Thus, supercomputing as a platform for accomplishment can and is resourced through mutual engagement even as individual accomplishment is realized through initiative and enterprise.

The authors make clear the most important message through a series of well-structured presentations from a diversity of sources: supercomputing is ubiquitous, is being rapidly embraced across an ever-expanding breadth of industrial and commercial regimes, and as a result is emergent as the catalyst for economic growth and sustainable vitality in the new industrial age. The challenge to all is in education. Once limited to a few apprenticeships, education in high-performance computing is finding new channels of dissemination and presentation across the world. This book provides both breadth and depth of examples across disciplines and nations as to what is being accomplished and what may yet be achieved. You will find this an informative and delightful read. Enjoy and be inspired.

Thomas Sterling
Center for Research in Extreme Scale Computing
School of Informatics and Computing
Indiana University Bloomington, Indiana, USA

Acknowledgments

Many people in the supercomputing business have contributed to this book. Some of them have been in this business far longer than we have, and they have provided valuable historical context for Chapter 1. Bill Kramer, NCSA's Director of the Blue Waters Project, nearly overwhelmed us with computing history books, but he had them readily available and we read every one of them. Beth McKown, also from NCSA, appeared one day with industrial letters of testimony that were surely hidden in a basement archive since 1995, but she knew where they were, and, more importantly, that they would add insight in a way that may never previously have been shared publicly. Hervé Mouren from France graciously counseled us at a Courtyard Marriott in Paris when we were in self-imposed exile to create book strategy and content. He has inspired us to further document the cultural and national reasons for specific-country competencies and behavior. Additionally, the German and British legacy on historical information related to World War II efforts on things which eventually led to the computing business as we know it now have been instrumental to illustrate the effort to enter the digital era.

On various occasions, we have discussed and acknowledged the cultural and political challenges to industrial supercomputing that are shared between Europe and the United States of America, which ultimately led to both of us endeavoring to write a book. It has turned out that writing and editing has been made easier since our vision for the book was borne out of a need to reach a non-technical audience while not embarrassing ourselves among our technical colleagues.

Michael Resch (Germany), Sang Min Lee (South Korea), and Hiroshi Kawai (Japan) have willingly kept the faith regarding industrial supercomputing and have each pledged to continue the good work of our International Industrial Supercomputing Workshop. Your support has been both real and genuine.

Dona Crawford, who co-chairs the U.S. Council on Competitiveness HPC Advisory Committee and whose day-job is at Lawrence Livermore National Lab in California, kindly agreed to add authorship and editing to her already busy schedule. Without her, our readers would have very little insight into the outreach of key federal supercomputing laboratories.

NCSA's Private Sector Program partners and staff bought into the idea of a contributed book, which would actually require them to provide their own contributions. Thanks, team!

From Europe, we have, among others, relied on many of our friends in the PRACE partnership for advanced computing in Europe. Very valuable information and cases have been brought to us, which we have gratefully used in this book. European colleagues like Francesc Subirada (BSC), Stéphane Requena (Genci), Andreas Wierse (HLRS), Claudio Arlandini (Cineca), Peter Michielse (SURFsara), and also Augusto Burgueno-Arjona (European Commission—who kept insisting on linking PRACE with industry) have directly or indirectly supported or inspired us in finishing our endeavor. Their role was very important for us.

The most important acknowledgment is for our wives, Erika and Anke, who have endured our solid hours and days at the computer patiently, quietly, and kindly. Some of these hours were during family time and some were well after midnight. Our endless gratitude goes out to you!

And finally, we wish to acknowledge all of you in the supercomputing business who have encouraged us to finish this book, and even dare us to pursue a sequel. Your encouraging words are helpful—especially as you suggest that no book like this has ever been written. Please buy copies for yourselves and others, and continue to share with us your ideas of how to bring HPC closer to non-technical readers.

Warmest regards,

Merle Giles and Anwar Osseyran

Executive Summary

THE VISION FOR THIS BOOK BEGAN OVER DINNER IN AMSTERDAM following SURFsara's Superdag annual event for The Netherlands HPC community to get informed on recent developments and opportunities, and get in touch with each other. Superdag is specifically organized for users of the national ICT (Information, Communication, and Technology) research infrastructure at SURFsara. At this particular meeting, Prof. Anwar Osseyran invited Merle Giles from NCSA to speak about why we as a community should care about industrial supercomputing. The dinner request to collaborate on a book seemed a natural one, as the authors had met previously as part of an informal group made up of supercomputer centers that serve and collaborate with industrial users.

This informal group was named International Industrial Supercomputing Workshop, an appropriately descriptive tag first used by Dr. Sang Min Lee at KISTI Supercomputer Center in Daejeon, South Korea, in 2010. The second workshop was hosted by Dr. Michael Resch at HLRS in Stuttgart, Germany, the home of world-class collaboration with automobile manufacturers Mercedes Benz and Porsche. The third was hosted at NCSA at the University of Illinois, the fourth at SURFsara in Amsterdam, and the fifth by University of Tokyo in Japan.

The rules were simple: attendees must come from supercomputer centers that serve industrial users, and vendors were not allowed. Typical supercomputer meetings are rather heavy on technology, as this business is fast-moving and very innovative. Industry, in this case, was defined as end-users of supercomputing technology rather than the more common reference of supercomputing as a provider industry. The common bonds of this group were rooted in the customer service models and unique challenges of serving users from industry rather than university and academic researchers. A fundamental difference between these user communities is that academic researchers obtain allocations on supercomputers as part of peer-reviewed grant processes, and they run applications on these machines at no cost. Industry, on the other hand, regularly pays for such use.

Historically, supercomputers are funded primarily by (1) national governments, (2) universities, and (3) individual companies. These resources differ mostly in terms of scale, with the largest computers funded mostly by nations, mid-tier machines by universities, and mid-tier to entry-level computers by individual companies. Social and economic benefits occur along all scales, but the best-known and most publicized benefits are from academic communities, which by virtue of receiving no-cost allocations must publish their results for the greater good. Industry use ranges from pure commercial and production purposes to precompetitive and open scientific use, with the latter requiring the sharing and publishing of results.

We attempt to tell a story about how industry demand for supercomputing is rooted in science and engineering that is similar to that of academic science and engineering. Differences occur in the scale of discovery and knowledge, with academic scientists seeking scientific insight, whereas industrial scientists and engineers seek insight that manifests in products and services, typically on a shorter time-to-discovery scale. Big science can use large fractions of supercomputers, whereas engineering typically uses smaller fractions. Academic research allocations on supercomputers may lead to publishable discoveries with no guarantee of subsequent allocations, whereas industrial use tends to run iterations of simulations to seek an optimal solution. The former can mostly be described as capability computing when the researcher makes optimal use of the machine capabilities to solve one large complex problem; the latter is often described as capacity computing. We should point out that scientists in both academia and industry require capability computing for scientific insight and that both communities utilize capacity computing routinely. The distinction here is that government policies around the world often favor capability computing in the pursuit of science; private-sector computers favor optimization and capacity computing. In all cases, the terms supercomputing and high-performance computing, or HPC, will be used interchangeably. The Supercomputing 101 introduction will help nontechnical readers understand the fundamental aspects of how HPC differs from normal, everyday computing.

This book is not intended to be a comprehensive review of the global state of supercomputing. It does attempt to give the reader a global sense of technological purpose. Provisioning vast technical resources, operating these systems, programming them, and taming these fickle beasts take immense resources for personnel. Because they are not mass-market machines, procurement costs are high, and deployment is risky. Once tamed, the technical aspects of these super slick computers become commoditized and costs decrease. An extreme example of this is that, according to Electronista's interview with Jack Dongarra, co-manager of supercomputing's TOP500 List, Apple's iPad 2 in 2010 is as fast as a 1985-era Cray 2 supercomputer. Moreover, the Cray 2 was the size of a washing machine.

Technical and nontechnical readers will benefit from Chapter 1. It is written in a way to be chronological, with perspective from multiple countries. A careful reader will learn that the initial urgency for computing during World War II in the 1940s was identical in Japan, Germany, and the United States. A persistent need for speed split the computing family tree in the 1960s between business computing and supercomputing, the former becoming classical IT, or information technology, and the latter building the support for the world's most intractable science and engineering.

Authors from 11 countries contributed to this book, describing their supercomputing environments in terms of policy and service models, and many have shared industry case studies. The uniqueness of these countries comes through their stories, and their similarities will become obvious with careful observation. The reader will learn that the providers of supercomputers are few, yet the users, both academic and industrial, are numerous. It is our hope that the reader will gain an awe-inspiring appreciation for the world's computing pioneers, as these engineers and entrepreneurs created an entirely new industry that is central to society as we know it. It may be less obvious as the case studies fail to represent the sum of all industry, but we also hope to present the notion that supercomputing is necessary to foster healthy economies by positively affecting product quality while lowering costs, and that some physical tests are impossible to achieve given the complexity of modern manufacturing.

More subtle and technical aspects of industrial supercomputing will be found in the case study Chapters 15 through 23, while smaller case studies and descriptions will be found in the country chapters. Europe, Asia, and the United States are represented, but no attempt has been made to be comprehensive. The supercomputer centers in developed countries, which contributed chapters, do not speak for their entire country. Indeed, these countries often have more than one supercomputer center. The representation of industrial supercomputing will certainly be incomplete, but we hope that this book will inspire others to share their stories in different ways, and that policy makers and industrial decision makers might be more informed about the economic impact of such powerful technological investments.

As authors and coeditors, perhaps we have learned the most. It is our sincerest desire to share some of these learnings with you, whether you are a technical neophyte or a computer science genius. At the very least, we learned that business models matter, and that all solutions are not technical. Very real barriers exist to success, including cultural, political, and business practices. At most, we expect the reader to appreciate that there is no going back. Technical advances through industrial supercomputing have become a lifestyle for many companies, and continued gains in productivity and quality may usher in a digital industrial revolution of equal or greater magnitude than the first.

It is clear to us that in this era of global financial uncertainty, and occasional decline, supercomputing is a powerful, modern scientific instrument. The historical use of such instruments has changed the wealth of nations. If the space race was competitive, so is supercomputing, yet we hope to show that investment returns exist at significant multiples, and that there is an increasing need for collaboration as companies, national agencies, and entire countries may not be prepared for the scope of investment required to do it right. We also hope to show that centralized supercomputing at a scale that matters is a powerful key to success for companies, as well as nations, large and small.

There are certain challenges of science that, until they reach a certain threshold of scale to contain the entire problem, make every attempt at lesser scale irrelevant. Large-scale supercomputers, therefore, are simply the *only* way to learn certain truths. Three diverse examples include star formation, stress testing macroeconomies, and determining the way in which water freezes at the molecular level. It is our observation that access to supercomputing matters, and that science and engineering problems at great scale and complexity exist in corners of the planet that are not well documented. The NDEMC case study in Chapter 21 describes how extreme multiphysics challenges can come from small companies in sparsely populated areas. The GE case study in Chapter 19 shows how complexity occurs at molecular levels in ways that cannot be adequately analyzed in a traditional laboratory. Indeed, industry-induced innovation is pushing the boundaries of known science, engineering, computing, and software. Industry-induced innovation will drive research in ways that can only be answered by equally digitally-capable scientists.

For these reasons, the conclusion chapter will suggest that collaboration is more important than ever. Perhaps this collaboration is precompetitive among companies in single sectors. Perhaps collaboration will be needed between noncompeting sectors and between sectors that do not naturally meet face-to-face. Collaboration may even be needed between political organizations, or government agencies, or entire governments as we see in Europe. Dare we suggest that intercontinental collaboration may even be needed?

It is our final hope that this book will aid in the understanding that collaboration is especially valuable based on the belief that national productivity and wealth are inextricably linked to industrial productivity and wealth. Recognizing the benefits of shared investments in supercomputing and the codependencies between public and private sectors, as well as among nations, places a burden and an opportunity on the shoulders of the supercomputing community.

We are optimists and realists. We trust that the passion and vision of our dear friends and colleagues around the globe will shine through and inspire you as it has us.

Anwar Osseyran
University of Amsterdam and SURFsara B.V.
Merle Giles
National Center for Supercomputing Applications (NCSA), University of Illinois at Urbana-Champaign

Contributors

Claudio Arlandini
Supercomputing Application and Innovation Department
CINECA
Segrate, Milano, Italy

Rick Arthur
GE Global Research
Niskayuna, New York, USA

Sanzio Bassini
Supercomputing Application and Innovation Department
CINECA
Casalecchio di Reno, Bologna, Italy

Serge Bogaerts
CENAERO
Gosselies, Belgium

Evan Burness
National Center for Supercomputing Applications (NCSA)
University of Illinois at Urbana-Champaign
Urbana, Illinois, USA

Steve Conway
High-Performance Computing/Data Analysis and Steering Committee Member
HPC User Forum
Minneapolis, Minnesota, USA

Dona L. Crawford
Lawrence Livermore National Laboratory
Livermore, California, USA

Chirag Dekate
DataDirect Networks Inc.
Boston, Massachusetts, USA

Merle Giles
National Center for Supercomputing Applications (NCSA)
University of Illinois at Urbana-Champaign
Urbana, Illinois, USA

Keith Gray
BP
Houston, Texas, USA

Mauricio Hanzich
Barcelona Supercomputing Center
Barcelona, Spain

Erwan Jacquin
HydrOcean
Nantes, France

Anni Jakobsson
CSC – IT Center for Science Ltd.
Espoo, Finland

Seid Koric
National Center for Supercomputing Applications (NCSA)
University of Illinois at Urbana-Champaign
Urbana, Illinois, USA

Sang Min Lee
Korea Institute of Science & Technology Information
Daejeon, South Korea

Cynthia McIntyre
Strategic Operations & HPC Initiatives
Council on Competitiveness
Washington, D.C., USA

Peter Michielse
SURFsara B.V.
Amsterdam, The Netherlands

Francisco Ortigosa
Repsol
Woodlands, Texas, USA

Anwar Osseyran
University of Amsterdam and SURFsara B.V.
Amsterdam, The Netherlands

Per Öster
CSC – IT Center for Science Ltd.
Espoo, Finland

Marc Pariente
Renault
Paris, France

Mark Parsons
EPCC
The University of Edinburgh
Edinburgh, United Kingdom

Andrea Penza
Supercomputing Application and Innovation Department
CINECA
Segrate, Milano, Italy

Raffaele Ponzini
Supercomputing Application and Innovation Department
CINECA
Segrate, Milano, Italy

Josep de la Puente
Barcelona Supercomputing Center
Barcelona, Spain

Stéphane Requena
GENCI
Paris, France

Michael M. Resch
University of Stuttgart High Performance Computing Center
Stuttgart, Germany

Catherine Rivière
GENCI
Paris, France

Todd Simons
Rolls-Royce Corporation
Indianapolis, Indiana, USA

Dane Skow
Dane Skow Enterprises
Winfield, Illinois, USA

Francesc Subirada
Barcelona Supercomputing Center
Barcelona, Spain

Yves Tourbier
Renault
Paris, France

Makoto Tsubokura
Division of Mechanical and Space Engineering
Graduate School of Engineering
Hokkaido University
Hokkaido, Japan

Ari Turunen
CSC – IT Center for Science Ltd.
Espoo, Finland

Andreas Wierse
SICOS BW GmbH
Stuttgart, Germany

Jeffrey P. Wolf
GoToMarket Consulting
Livermore, California, USA

Francis Wray
EPCC
The University of Edinburgh
Edinburgh, United Kingdom

Masako Yamada
GE Global Research
One Research Circle
Niskayuna, New York, USA

CHAPTER 1

Introduction and Brief History of Supercomputing

Merle Giles and Anwar Osseyran

CONTENTS

1.1 SUPERCOMPUTING 101

It should be noted that supercomputing, high-performance computing (HPC), and advanced computing are terms that are often used interchangeably. These advanced machines use components similar to everyday computers, such as memory, processors known as CPUs (central processing units), data storage, communication, and software, but supercomputers use these in ways that produce speedy results. The "trick" to supercomputing is in the scale of use by stringing together vast numbers of each of the components listed above and parallelizing data movement at every possible step.

Typically, communication between multiple computers in a data center occurs using common Ethernet cables, often seen in the form of orange or blue cables in homes and businesses. In supercomputing applications, even faster communication is used and highly proprietary communication schemes push speeds to 4–100 times the speeds of common Ethernet, using parallel, simultaneous data paths to push and pull data from CPUs to storage and memory and back.

Four basic components of supercomputing can be described as follows:

1. *Parallel processing*, using multiple cores per CPU, multiple CPUs in a single computer, and multiple computers in a supercomputer. A good consumer CPU example is a dual-core or quad-core processor in a laptop computer, which allows software applications to do two or four operations at once. CPUs used in supercomputers will typically have between 8 and 32 cores, with newer general purpose graphics processing units (GPUs) deploying hundreds of cores.

2. *Large memory*, with each computer holding 32, 64, 256, or more gigabytes of memory, depending on the specific science or engineering need.

3. Multiprocessor, or *multinode communication*, allowing an entire collection of computers to be used in a single application. Special software techniques, such as MPI (Message Passing Interface) must be used for sharing an application across two or more computers. Supercomputers might pass information simultaneously among 25,000 or more individual computers.

4. *Parallel file systems* move data between compute nodes and storage (disks, magnetic tape, and solid-state devices), using multiple channels simultaneously, rather than one, to speed up data input and output. In very advanced systems, large caches of memory sit between the storage and compute systems to minimize bottlenecks.

As in all computers, supercomputers require an operating system—commonly Linux—as well as machine-management software known as middleware. Specific application software will map to the operating system, and ultimately to the machine components themselves through the use of compilers, which translate word-based programs into machine language that only recognizes zeros and ones.

Supercomputing adds significant complexity for users due to idiosyncrasies of various vendor platforms, processors, storage disks, and so on. These differences are routinely subtle and nonobvious, leading to difficulties in fully understanding and exploiting myriad combinations of capabilities. Extreme scale requires incredibly complex paths through which to move data and communicate among tens of thousands of nodes, some of which may be in the same rack, or an adjacent rack, or the 200th rack 100 ft away. The complexity comes from trying to minimize the number of hops from node to node, and the latency, or time, to travel to and from a distant computer. Less latency makes for faster computing.

Data storage can be installed in large quantities, both in terms of aggregate storage and in how much data can be packed into a single unit. The largest data storage systems in the world can hold nearly 400 petabytes of data, or 400,000 terabytes. Modern desktop computers hold one, two, or three terabytes of data on disk.

Large digital numbers for storage capacity use the following system of names:*

1 byte = 1 computer word

1000 bytes = 1 kilobyte (KB), or 1000^1 bytes (10^3)

1000 kilobytes = 1 megabyte (MB), or 1000^2 bytes (10^6)

1000 megabytes = 1 gigabyte (GB), or 1000^3 bytes (10^9)

1000 gigabytes = 1 terabyte (TB), or 1000^4 bytes (10^{12})

1000 terabytes = 1 petabyte (PB), or 1000^5 bytes (10^{15})

1000 petabytes = 1 exabyte (EB), or 1000^6 bytes (10^{18})

1000 exabytes = 1 zettabyte (ZB), or 1000^7 bytes (10^{21})

1000 zettabytes = 1 yottabyte (YB), or 1000^8 bytes (10^{24})

The same nomenclature is used for memory, such as "64 gigabytes of memory," and for floating point operations per second (FLOPS). Aggregating these terms, today's supercomputers are of the petascale generation, and the generation desired in the next 10 years will be known as exascale.

---

* Throughout this book, we have chosen to use the decimal system of prefixes to bytes and flops. We realize that for solid-state memory, the binary method is more common (1024 instead of 1000), whereas for disk sizes, transmission speeds, and calculations, the decimal system (1000) is used. See also http://en.wikipedia.org/wiki/Wikipedia:Manual_of_Style/Dates_and_numbers#Quantities_of_bytes_and_bits.

1.2 THE DAWN OF DIGITAL ELECTRONICS: 1930s–1940s

According to *A History of Computing in the Twentieth Century*,* the first major computing activity in America took place around 1935 in Cambridge, Massachusetts, in the form of an analog machine developed by Vannevar Bush, called the differential analyzer. Analog machines were either mechanical or made of a variety of vacuum tubes, and were originally developed for radio and telephone communications.

In Japan, in 1938, Mr. Shiokawa Shinsuke of Fuji Electric invented a relay binary circuit. Together with Dr. Ono, professor of mathematics at Tokyo University, they conceived the idea of using a binary system in a statistical computer. This statistical relay computer was ultimately completed in 1951 with the cooperation of Dr. Yamashita, professor of electrical engineering at Tokyo University, who learned that relay computers were more reliable than vacuum-tube computers. This 13-year delay was due to the war effort, which stopped research and also created a shortage of materials and American dollars, making it difficult to import critical raw materials and measuring devices. Perhaps due to these lack of materials, the 1950s spurred proprietary Japanese design.

As a young man in Czechoslovakia unaware of computers, Antonin Svoboda believed he would become a theoretical physicist. In 1937, however, he met fellow Czechoslovakian Dr. Vladimir Vand and began working for the war effort. Vand learned of mechanical differential analyzers, and added one parameter to the system—solving for a single variable, then its first and second derivatives. This innovation allowed Vand and Svoboda to design an antiaircraft gun control system for the Czechoslovakian Ministry of National Defense. Svoboda later escaped the war in Europe and ultimately retired from University of California at Los Angeles.

Konrad Zuse, a civil engineering student in Berlin, Germany, in 1934 and later an engineer in the aircraft industry had been aware of the tremendous number of monotonous calculations for the design of static aerodynamic structures. Svoboda ultimately designed and constructed calculating machines to solve these problems automatically, resulting in a fully operational model dubbed Z3 in 1941.

In the Union of Soviet Socialist Republics (USSR), World War II also interrupted a series of scientific research and technical projects in the field of mathematical machinery, as it was called. There existed a strong tradition of interest in applied mathematics and calculation methods in the

* The Lax Report can be downloaded in its entirety at this link www.cray.com.

USSR, but war focused those efforts on research into numerical methods and automation of calculations applied mainly for control of artillery fire.

Arguably, the first working, modern, electronic, digital computer was the Colossus machine, put into operation in 1943 at the United Kingdom's Government Code and Cypher School (CG & CS) in Bletchley Park. Although it was designed to break a specific German cipher, the Colossus was a true electronic computer that could be used, in principle, on a wide range of problems. Colossus was designed by Tommy Flowers, whereas Alan Turing's efforts in breaking the German cipher (ENIGMA) contributed to the design of Colossus. Its existence was classified until the 1970s.

America's development in computing was accelerated during the war, with U.S./U.K. collaboration playing a major role in creating the early U.S. computer industry. In particular, U.S. engineers at a National Cash Register plant in Dayton, Ohio (deputized into the war effort), built copies and improved versions of Bletchley Park electronic cryptanalysis machines, as well as computers of their own design. Thus, the United States' development of computer technology was directly linked to the need for computing speed and accuracy, driven by cryptology, nuclear weapons, and antiaircraft firing controls.

More specifically, America entered into the computer field through the initiative of Colonel Leslie Simon from the U.S. Army's Ballistic Research Laboratory at Aberdeen Proving Ground, Maryland. As Germany invaded Poland in 1939, Colonel Simon approached RCA (founded as the Radio Corporation of America) about how to improve an antiaircraft gunner's chances of shooting down enemy planes. To hit targets on the ground, a gunner could make use of firing tables prepared in advance. But to place a shell in the path of a moving airplane required on-the-spot computation, including last-minute estimates of time-to-flight to set a timed fuse so the shell exploded as close to the target airplane as possible. And while guns could shoot, it was very difficult to aim with the mechanical "directors" of the day, which were too slow. Colonel Simon had the foresight to believe that electronics could provide the required speed while in the air.

Vacuum tubes were proposed for a revolutionary machine that could compute ballistic trajectories at least 10 times faster than a *mechanical* differential analyzer, and at least 100 times faster than a human computer with a desk calculator. (The original use of the word "computer" began in the 17th century and referred to a person performing mathematical calculations manually.) One estimate of the day was that a human computer could calculate a single trajectory in about 12 hours, with hundreds of trajectories required to produce a firing table for any particular combination of shell and gun. The *electromechanical* differential analyzer could calculate a trajectory in 10 or 20 minutes and a complete firing table in about 1 month. It soon became clear at the Army Ballistic Research Laboratory, however, that no amount of augmentation of the staff of human computers would suffice, thus providing the rationale to sponsor the development of a radically new machine, ultimately named the Electronic Numerical Integrator and Computer, or ENIAC. While ENIAC's immediate goal was to calculate firing tables, its long-range goal was to be a general-purpose computer. ENIAC began to solve its first problem in December 1945 for Los Alamos National Laboratory (LANL) in New Mexico. The problem was of such importance that details of this work remain classified at least until the 1980s.

Central to this electronic revolution was Whirlwind I, an early ColdWar vacuum-tube computer developed by the MIT (United States' Massachusetts Institute of Technology) Servomechanisms Laboratory for the U.S. Navy. It was among the first digital electronic computers that operated in real-time for output, and the first that was not simply an electronic replacement of older mechanical systems. In 1945, Perry Crawford of MIT saw a prototype of ENIAC and suggested that a digital computer would allow the accuracy of a simulation to be improved with the addition of more code in the computer program, as opposed to adding parts to the machine. As long as the machine was fast enough, there was no theoretical limit to the complexity of the simulation.

Up to this point, all computers were dedicated to single tasks and ran in batchmode. That is, a series of inputs were fed into the computer, which would work out the answers and print them. This was not appropriate for Whirlwind, which needed to operate continuously on an ever-changing series of inputs. Speed then became a limiting issue. Although with other systems it simply meant waiting longer for the printout, with Whirlwind it meant significantly limiting the amount of complexity the simulation could include.

It soon became apparent that digital computers would be much more useful for the American war effort for solving urgent computational problems such as calculating ballistic tables than for fire control, or trajectory tables. (Ballistics is the science of mechanics that deals with the launching, flight, behavior, and effects of projectiles, especially bullets and rockets.) The very large number of tubes required made them impractical for use on airplanes, for which their speed, though advancing rapidly, was still insufficient for real-time computations. On the other hand, their three-orders-of-magnitude-greater speed with respect to mechanical computers made them a godsend for lengthy numerical computations.

Bomb making and ballistics needed great strides in computing capability, so LANL continued its work in ballistics following the war. Meanwhile, Edward Teller started another U.S. government laboratory for the purpose of building a hydrogen bomb. Of note, the unique defining characteristic for what would become known as Livermore Branch of the University of California Radiation Laboratory would be electronic digital computing.

Late in 1945, Hungarian-born John von Neumann, who came to America's Princeton University (New Jersey) as a professor in 1933, decided to build a machine at Princeton's Institute of Advanced Study (IAS) based on an electrostatic storage tube to be developed at RCA Labs. One rendition of von Neumann's machine was named JOHNNIAC and was housed at the RAND Corporation. In 1949, Argonne National Laboratory (ANL) decided to have an IAS computer built to facilitate the computations related to nuclear reactor design, to be named AVIDAC (Argonne's Version of the Institute's Digital Automatic Computer). Construction began in 1950. That same year, Oak Ridge National Laboratory (ORNL) decided to build its own, which would be the brother to AVIDAC, named ORACLE (Oak Ridge Automatic Computer and Logical Engine).

EDSAC (Electronic Delay Storage Automatic Calculator) was the world's first full-size stored-program computer, built at the University of Cambridge, United Kingdom, by Maurice Wilkes and others. EDSAC was built according to the von Neumann machine principles established at IAS and, like the Manchester Mark I, became operational in 1949. Wilkes built the machine chiefly to study computer programming issues, which he realized would become as important as the hardware details. Wilkes' inspiration from *Project Whirlwind* led to the founding of MITRE (arguably an acronym for Massachusetts Institute of Technology Research Establishment, although MITRE's website does not acknowledge this) in 1958 to provide overall direction to the companies and workers involved in the U.S. Air Force SAGE (Semi-Automatic Ground Environment) project.[2] After the SAGE project ended in the early 1960s, the Federal Aviation Administration (FAA) selected MITRE to develop a similar system to provide automated air traffic control, further leading to development of the U.S. military Command, Control, Communications, and Intelligence (C3I) projects, including the Airborne Warning and Control System (AWACS). MITRE also worked on a number of projects with ARPA

(Advanced Research Projects Agency, renamed DARPA in 1972 by adding Defense) including precursors to ARPANET. Since the 1960s, MITRE has developed or supported most U.S. Department of Defense early warning and communications projects.

Meanwhile, an agreement in 1949 was made between the Army Ballistic Research Laboratories and the University of Illinois to construct two computers: ORDVAC and ILLIAC. ORDVAC was completed in November 1951 and was moved to Aberdeen in February 1952. ILLIAC was completed in September 1952, remaining in service until 1962. The 6-months before ILLIAC was born led to an early experiment with what would now be called a remote terminal. Terminals in Aberdeen and Urbana used the commercial Teletype network for reproduction on paper tape at each end.

Oddly, it appears that the dawn of digital electronics began in similar ways in Japan, Germany, and in the United States/United Kingdom so that we could shoot at each other more precisely while in fast-moving airplanes. National defense thus accelerated hardware design, programming, storage, and networking, including modern air traffic control systems.

1.3 THE DAWN OF SUPERCOMPUTING: 1950s–1970s

After graduating from university in 1951, America's Seymour Cray succeeded in leading development efforts of magnetic switches and transistors. Magnetic switches were used in an early Air Force computer and, by 1956, virtually every U.S. technology giant was dipping a toe into the digital water: Remington Rand, National Cash Register, Burroughs, IBM, Engineering Research Associates (ERA), General Electric (GE), RCA, Philco, Xerox, Bendix, Honeywell, Sperry Corporation, and Raytheon.

For computer manufacturers, the needs of the weapons labs created incredible opportunities. Throughout the early and mid-1950s, Livermore and Los Alamos stepped up their computing efforts and vied for prestige, funding, and access to the first of every kind of computer. In the late 1950s, the fastest and most expensive computers were three to four orders of magnitude more powerful than the smallest models sold in large numbers. By the early 1970s, the spread exceeded four orders of magnitude.[3]

Control Data Corporation (CDC) was founded in 1957 and introduced its 1604 computer in 1960. Its first buyer was the Naval Post-Graduate School in Monterey, California. CDC sold more 1604s to the University of Illinois, Northrop, Lockheed, and the National Bureau of Standards. The story of CDC's success spread quickly, bolstered by the story of its

35-year-old computer designer, Seymour Cray. Weapons builders at Los Alamos and Livermore looked to him for their next-generation machines, as did the cryptologists in Washington, D.C.

Entering the computer field in 1952, IBM's 701 (Defense Calculator) and follow-on machines were delivered to federal defense and nuclear labs in the United States, United Kingdom, and France, as well as to the U.S. Weather Bureau, Massachusetts Institute of Technology (MIT), MITRE Corporation, and Bank of America. Naturally, MIT, MITRE, and the weather bureau were serving the defense community.

These early examples of IBM computer usage illustrate a central fact of the rapid growth of the computer field, namely, "devices placed in the hands of users quickly led to new applications, new capacities, new speeds, new peripherals, and new software systems." Thomas Watson, IBM's president in 1955, was on the cover of *Time*, and his company's name was suddenly synonymous with computing—especially *business* computing. By 1960, IBM dominated the business market and scientific computing. Livermore bought 20 computers between 1954 and 1961, 19 of which were IBM. In 1961, Livermore bought a CDC 1604, breaking the string of IBMs.*

In the late 1960s, Cray planned to advance computer speed in two ways: (1) by cramming circuits closer together, and (2) by using four processors in parallel to replace the use of a single processor. And by 1972, Cray knew it was time to use integrated circuits, or chips, that were commercially available at the size of a baby's fingernail, incorporating transistors, diodes, resistors, and capacitors on a single chip of silicon.

On the CRAY-1, Cray took full advantage of the work of his predecessors. Los Alamos deemed the CRAY-1 to be five times faster than the previous CDC machine, and early customers lined up to buy. Early adopters were once again defense labs and weather researchers in the United States and United Kingdom, including Bell Laboratories and the National Security Agency. Automakers employed them for crash tests and air flow studies, aircraft builders used them to design wings and fuselages, and oil companies applied them to petroleum exploration.

America's history of computing is much like that of other leading nations in that early applications were primarily military ones. Postwar

* In 1963, CDC's total employees numbered 3500, up from 325 in 1959. By 1965 it had more than 9000 people worldwide and it had become the third largest computer maker in the world, trailing only IBM and the Univac Division of Sperry-Rand. Most of the world now recognized Control Data as the undisputed leader in the production of scientific machines. IBM, of course, was the industry heavyweight with 80% of the worldwide computer market.

1950s computers first took advantage of electronics and stored memory, which was the dawn of *business* computing. The 1960s introduced additional speed through parallel processors and by cramming circuits closer together, which fostered the first *supercomputers*. Thus, the differentiation for supercomputers was *speed*, brought on by parallelization and miniaturization.

The only serious competition for U.S. vendors came from the Japanese, who had enhanced vector-based supercomputing. In 1971, commercial availability of integrated circuits, or chips, led to Intel's first simple, single-chip processor (or microprocessor). The use of low-cost, mass-produced, high-volume commodity microprocessors was to transform all segments of the computer industry and put serious pressure on vector-based computing.

1.4 GROWTH AND COMMODITIZATION OF SUPERCOMPUTING: 1980s–2000s

In 1980, Japan announced a supercomputer project aimed at creating a machine capable of speeds at least three orders of magnitude higher than was available at the time, or roughly a 10-gigaflops computer. As a comparison, following are operating speeds and in-service dates of several existing and planned U.S. supercomputers:

- 1976—CRAY-1: 88 megaflops (million floating point operations per second)
- 1981—CDC Cyber 205: 400 megaflops
- 1984—CRAY-2: 800 megaflops

Six major Japanese vendors joined the National Super-Speed Computer Project (NSCP Project: Fujitsu, Hitachi, NEC, Mitsubishi, Oki, and Toshiba). Total funding for this project was estimated to be about U.S. $200 million.

In 1993, the *Beowulf Project*, based on NASA's requirement for low-cost workstations, leveraged PCs, Intel's integrated microprocessors, and highly available operating systems such as Linux and Windows to commoditize supercomputers.[4,5] This design competed against higher-cost vector machines and enabled do-it-yourself techies to further commoditize supercomputing and form communities. Training and education became more efficient as software would operate on more than a single vendor's machine. In 1994, a 16-node, $40,000 cluster was built from Intel 486 computers.[6] A Beowulf cluster won the Gordon Bell Prize for performance/price in 1997. By December 2000, 28 Beowulfs were in the TOP500 and the total population was estimated to be several thousand.

Select highlights of this era include the following[7]:

- 1983: ARPA initiates its Scalable Computing Initiative (SCI) in search of a parallel processing paradigm based on many low-cost microprocessors, generating the move to Massively Parallel Processing (MPP).
- 1985: The U.S. National Science Foundation (NSF) forms NSFNET, linking its five supercomputing sites.
- 1986: Intel's 80386 chip debuts on PCs, making graphical PC operations practical.
- 1990: Geneva's CERN develops World Wide Web URLs and HTML.
- 1991: NSF adds commercial users to its NSFNET; Finnish university student Linus Torvalds releases Linux, leveraging MIT's GNU to become completely open source.
- 1993: Mosaic UNIX web browser is released by NCSA's (National Center for Supercomputing Applications) Bina and Andreesen; Intel releases the Pentium, its fifth generation "×86" chip.
- 1994: Mosaic takes off after being made available on Windows and Mac OS, ultimately becoming Netscape Communications Corporation.
- 1995: Internet becomes self-supporting after NSFNET backbone decommissioned.
- 1995: ARPANET is renamed the "Internet."
- 1999: Supercomputing splits between Cray's vector computing and Massively Parallel Processing (MPP) computing, with Beowulf standardizing computer clusters with a common platform and programming model independent of proprietary processors and networks.

To be clear, microprocessor speeds found on desktops in 2014 had overtaken the power of supercomputers in the 1990s. Video games were using the sort of processing power previously available only to government laboratories. The first U.S. Bush administration in 1990 had defined supercomputers as being able to perform more than 195 million theoretical operations per second (195 MTOPS), but by 1997 ordinary microprocessors were capable of 450 MTOPS.[8]

The reader should also take careful note that the Beowulf Project commoditized supercomputing and created a split between vector-based computing and MPP. Commoditization increased access, as did graphics-based Windows systems, driving broad access to the Internet for scientists and consumers. The modern digital era was thus borne out of defense-driven supercomputing and networking.

1.5 SUPERCOMPUTERS FOR OPEN SCIENCE AND OPEN ENGINEERING

At the end of World War II, scientists were highly regarded for making major contributions to the war effort, such as development of the atomic bomb, penicillin, and radar. Trivia: rapid commercial production of penicillin was required by those wounded in war and was made possible by using lactose liquid, a by-product of producing corn starch at the Northern Regional Research Lab in Peoria, Illinois, to accelerate penicillin production tenfold.* These had a profound effect on the outcome of the war. The reader will recall that Vannevar Bush was a leading engineer and inventor of the differential analyzer in Cambridge, Massachusetts, around 1935.

At the beginning of the war, Bush convinced U.S. President Franklin D. Roosevelt to make use of science, engineering, and technology capability to help the war effort. Near the end of the war, Roosevelt asked Bush to detail ways to harness science and engineering talent after the war. Bush and others began preparing a report that was titled *Science—The Endless Frontier.* It was a plan for funding university researchers and science education by establishing a new federal agency. While the report was being prepared, Roosevelt died and Harry Truman became President. Truman allowed the report to be made public, which led to a congressional bill being signed in 1950 to set up the NSF. This established a national policy for the promotion of "basic research and education in the sciences," with engineering added as a priority in 1981. It is the NSF that ultimately launched America's seminal investments in open-research supercomputing.[9]

In the 1970s, Dr. Larry Smarr, an American astronomy professor, had been traveling to West Germany to use supercomputers for astrophysics

* http://www.lib.niu.edu/2001/iht810139.html.

research. His German collaborators teased him about the irony of an American researcher forced to come to a nation still filled with American troops just to use supercomputers that had been designed and built in the United States in the first place. Other American scientists faced similar situations. They built their research around national security concerns and went through the rigors of getting clearance on Department of Energy (DOE) machines. Or they went to Europe, where time and access were more readily available, and they made do with smaller machines with significantly less power.

Dr. Smarr and seven of his faculty colleagues from the University of Illinois at Urbana-Champaign (John Kogut, David Kuck, Robert Wilhelmson, Peter Wolynes, Karl Hess, Thomas Hanratty, and Robert McMeeking) responded to this "famine" of American supercomputing with an unsolicited proposal to the NSF asking for funding to launch open-research supercomputing centers. This "Black Proposal," officially titled "A Center for Scientific and Engineering Supercomputing,"[10] claimed "several orders of magnitude improvement in computer power is urgently needed in American universities." The proposed center was meant to be a "model for new type of dedicated basic research facility in which large-scale problems of heretofore intractable complexity, in a wide variety of disciplines, could be successfully attacked." (The reader will recall early University of Illinois involvement with ORDVAC and ILLIAC in 1952.)

Included in the Black Proposal was a World Census of Supercomputers taken from a multiagency "Report of the Panel on Large Scale Computing in Science and Engineering," dated December 26, 1982, of which Peter D. Lax from New York University and the National Science Board was Chairman. The DOD, DOE, NASA, and NSF were sponsors and/or collaborators. In this report, dubbed the Lax Report, 60 machines were listed as a "Partial Inventory and Announced Orders of Class VI Machines," of which 43 were Cray-1s and 17 were CDC Cyber 205s:

41—United States

7—England

6—Germany

4—France

2—Japan

These Class VI supercomputers were defined as those with a memory capacity of at least 8 Mbytes,* a speed of upwards of 100 megaflops, and both scalar and vector registers in the CPU.† The 60 can be categorized as follows:

15—Weapons research, military, intelligence (14 United States; 1 England)

13—Open research (4 United States; 2 England; 1 France; 4 Germany; 2 Japan)

8—Physics and engineering (6 United States; 1 England; 1 Germany)

7—Petroleum (5 United States; 1 England; 1 Germany)

6—Atmospheric/Oceanographic/Weather/Environmental (4 United States; 2 England)

4—Nuclear and reactor research (3 United States; 1 France)

4—Timesharing (3 United States; 1 France)

3—Energy and Power (2 United States; 1 France)

The census reveals early adoption by industry, with 12 American supercomputers in private companies (42% of which were for petroleum research). A subsequent 1995 NSF report states that a program office was initially set up to purchase "cycles" from existing sources and to distribute those cycles to NSF research directorates. At least one of the original centers from which "cycles" were purchased was Boeing Computing Services, which is included in the three U.S timesharing centers listed above. The other centers at that time were Purdue University in Indiana and University of Minnesota.

Interestingly, while more than two-thirds of the world's supercomputers in 1982 were hosted in the United States, the "famine" referred to by Smarr was for open research, an area in which the Americans trailed the Germans. Note that the 1982 census simply counts machines and makes no distinction of relative computing power. Worldwide, only 8 of 60 supercomputers were hosted by universities.

As of mid-1983, the number of Class VI supercomputers in the world grew to approximately 75, yet Smarr claimed that the computing power

* On Cray systems memory was quantified with the term Word, which equals 8 bytes. So the memory capacity of 8 MBytes is equivalent to 1 MWord.

† http://onlinelibrary.wiley.com/doi/10.1029/EO064i049p00964-01/abstract.

drought in American universities resulted in a brain drain to weapons labs, industry, and foreign computing centers. Computational astrophysicists had been going to the Max-Planck Institute of Physics (MPIP) in Munich to do basic research. He further claimed that even mission-oriented applied researchers at DOE labs sometimes got more CRAY time at Germany's MPIP than at either Livermore or Los Alamos labs in the United States.

Japanese supercomputers from Hitachi and Fujitsu soon rivaled the power of the CRAY and CDC machines, furthering the notion that Japan and Germany (expecting a CRAY X-NP) would rapidly begin to take over leadership in scientific and engineering research from American universities. New CRAY-1 systems were installed at the French Ecole Polytechnique and various campuses of the University of Paris, making Europe one of the largest markets for supercomputer systems that year.

Bolstering the Black Proposal vision was a statement by University of Michigan's Brice Carnahan in the 1982 Lax Report:[11] "To date, Class VI machines have not been available to a broad spectrum of engineering researchers, particularly academic researchers. Despite this, these machines have already had a significant impact on research in several engineering disciplines. The problems tackled so far are characterized less by the particular discipline than by the mathematical formulations involved." This remains true today, with a continuing need for speed from supercomputing to simultaneously solve millions and billions of algebraic and differential equations.

NSF responded favorably to the Black Proposal in 1983 by establishing an Office for Advanced Scientific Computing. In 1985, NSF provided $48,050,000 in funding[12] for 5 years at four university centers, with a fifth center added in 1986:

1. NCSA at University of Illinois at Urbana-Champaign, at which Larry Smarr became its first director
2. Cornell Theory Center (CTC) at Cornell University, New York
3. San Diego Supercomputer Center (SDSC), University of California, San Diego, initially operated by General Atomics
4. Pittsburgh Supercomputing Center (PSC), directed by University of Pittsburgh and Carnegie Mellon University, Pennsylvania
5. John von Neuman Center, a consortium located at Princeton University, New Jersey (1986)

1.6 SUPERCOMPUTERS FOR INDUSTRY

Not surprisingly, investments in supercomputing technologies were viewed as critical to national competitive advantage, as recorded in a Congressional hearing on Federal Supercomputer Programs and Policies June 10, 1985.[13] Chairman Doug Walgren of the Congressional Subcommittee on Science, Research and Technology (which authorizes the NSF) stated that the federal government had supported the domestic computer industry since its inception, which stimulated the supercomputer industry by purchasing more than 50% of the nation's present-generation machines. He also shared that while supercomputing was a small part of the total computer market, their strategic value in advancing science and technology was quite significant, adding that to take full advantage of the potential contribution of the new national centers, special attention must be given to the use of peripheral services, development of national scientific networks, and research necessary to improve supercomputer software productivity.

Testifying at this Congressional hearing were supercomputer center directors from Illinois, Texas, Florida State University, Lawrence Livermore National Laboratory, representatives from the DOE, NSF, and DARPA, and the manager of engineering and product data systems at Ford Motor Company. The testimony from Ford Motor Company's Mr. Henry Zanardelli is most interesting and he cites two dominant factors for competitive advantage in the automobile industry: (1) product quality and (2) product cost, and that these factors are dependent to a great extent on the effectiveness of using supercomputers. Furthermore, the use of supercomputers for design evaluation usually involves three steps:

1. Development of a mathematical representation or structural model of the part, assembly system, or vehicle being designed, requiring the skills of a specially trained engineer.

2. Processing the model through computer simulation and analysis programs, requiring supercomputing to accomplish this in a reasonable amount of time.

3. Interpretation of the results by an engineer, at which point the achievement of design objectives is verified, or needed changes are identified.

We add a bit more historical perspective from this 1985 hearing, as it is striking how relevant the testimony is 30 years later. Zanardelli claimed that

Ford Motor Company was one of the earliest users of structural analysis known as finite element analysis (FEA), based primarily on software developed by the NASA, called NASA Structural Analysis, or NASTRAN. This was credited as an outstanding example of a government space program by-product that benefits the private sector. At that time, Ford was simulating large FEA models involving 15–20,000 discrete elements, with the physical property of each element defined in the model. Needed was a supercomputer capable of hundreds of millions of instructions per second.

Zanardelli goes on to claim cost reductions to one-fifteenth of the cost of a physical prototyping through the use of computer simulation by reducing the number of prototypes built, shortened time to develop new vehicles, improved product quality, and savings on designs and tests. As a comparison, today's advanced commercial FEA models can involve 30,000,000 elements or more, an increase of 1,500-fold over Ford's models from 1985. And sustained computing capacity on NCSA's Blue Waters in 2013 was more than one quadrillion calculations per second, a theoretical increase in computing power of more than 300,000-fold from mere hundreds of millions of calculations/second in 1985.

An important additional example of early industry success was published in the "Hayes" Report of 1995, named as such since the "Report of the Task Force on the Future of the NSF Supercomputer Centers Program" chairman was Edward Hays of The Ohio State University (read more on this in Chapter 13). The report named three kinds of industrial funding that have been significant to the NSF Centers program:

1. Funding from computer vendors, particularly suppliers of high-performance hardware and software
2. Funding from industrial users who wish to make use of the unique computational capabilities of the Centers
3. Funding from industries interested in technology transfer and training

Coterminous with NSF's investment in supercomputing was the formation of NASA's Numerical Aerodynamic Simulation (NAS) program in 1987 to focus resources on solving critical problems in aeroscience and related disciplines, as well as providing supercomputing power to solve computational fluid dynamics problems and to serve as a pathfinder in integrating supercomputing technologies, thus benefiting other supercomputing centers in government and industry. Project disciplines for NAS in the 1991–1992 operational year, for example, included aerodynamics, atmospheric sciences, hydrodynamics, hypersonics, physics, propulsion, structural mechanics, turbulence, chemistry, electrical engineering, engineering, geology, life sciences, and mathematics.

It has been said that no large manufacturer today can launch a product without using advanced modeling and simulation. Supercomputing simulations at both NASA and NSF centers achieve what cannot be done with physical prototyping or empirical methods alone—either the costs are too high, the pace too slow, the risks of error or harm to humans is too high or the experiment is simply impossible to realize. Likewise, modern Internet company adoption of HPC has swelled recently (see Chapter 2), resulting in greater than 52% of firms using supercomputer architectures to achieve valuable insights at speeds never before achieved, as reported by IDC (International Data Corporation), a global market research intelligence firm. The value propositions in the modern Internet sector range from quicker financial transactions to near-real-time fraud detection to personalized medicine.

1.7 MODERN INDUSTRIAL SUPERCOMPUTING

The increase in processor capabilities has been both a boon and a disruption. The 1980s and 1990s seemed to follow Moore's law in pure linear manner. In the 2000s, however, this linear increase in performance failed based on a fundamental inability to pack everything more tightly onto a single piece of silicon. Physical limitations prevailed, with more electrical current leaking than staying put. In response, chip vendors introduced more cores to the master processors, providing the ability to separate computer instructions into threads. This shift from seemingly endless increases in speed to forced threading, or parallelization of one's applications, has been disruptive.

Parallelization comes at the cost of rewriting software. In the video gaming industry, where the half-life of code is relatively short, this rewriting can generate significant returns in the marketplace. With legacy science and engineering codes; however, the costs to increase parallelism will be astounding. In 2009, a team from NCSA, Boeing, Caterpillar, General Electric, Procter & Gamble, and the U.S. Council on Competitiveness estimated that a multibillion-dollar software gap exists in creating and adapting application software to be capable of running on leadership-class supercomputers.

What began with single-processor integrated chips in the 1970s, mass-market commodity chips in the 1990s, and *multicore* processors

(2, 4, 8, 16, or more) in the 2000s more recently manifested in *manycore* processors with multiple tens or hundreds of cores on a CPU. GPUs, made famous by 3dfx, ATI, and NVIDIA, may have 1000s of cores on a single chip. These "advances" in hardware have placed an unusual burden on application programmers, as significant speed increases are only possible if applications are fundamentally rewritten.

Commodity clusters based on the integrated chip revolution in the 1980s now populate more than 80% of the TOP500 list. At a minimum, each of these clusters is using *multicore* processors. Supercomputers on the top of the TOP500 are now using *manycore* processors some in modest fractions and some completely. Companies, software developers, and perhaps entire nations must now pay close attention to the technical disruption caused by the manycore revolution in hardware.

One example of disruption comes from how Japan reached the modern era of supercomputing having been the earliest and most aggressive country in applying HPC to industry sectors such as automotive, aerospace, and chemical, but also consumer product design (rice cookers, etc., starting in the 1980s). When its peak performance on climate codes was first announced, NEC's Earth Simulator caused a Sputnik-like panic in some parts of the U.S. Congress and resulted in a funding boost for HPC in America. Japan then lost ground in HPC because it was slow to transition from vector technology to systems based on commodity processors. Some would say that a national strategy of funding three HPC vendors focused on vector processing in a single country was unsustainable. Japan's next-generation national supercomputer (K system) funding was decimated a few years ago and a bevy of Japanese Nobel laureates were placed into service to persuade government officials to restore most of the funding. NEC backed out of the project midstream, forcing Hitachi to withdraw and leaving Fujitsu as the last remaining company to complete the K supercomputer project. Funding was ultimately restored and Fujitsu and state-funded RIKEN finished the K in 2011 at a level of power greater than that of the next five systems on the TOP500 combined. Costs matter, as another high-profile example of disruption was IBM's unexpected 2011 termination of its 2008 contract to deliver a sustained-petaflops supercomputer at NCSA.

The decade following this book's publishing will certainly surprise and probably shock the supercomputing community. Cray, in a spectacular comeback move, ultimately won NCSA's NSF-funded Blue Waters award of $188 million, replacing IBM as the vendor. Probable shock was discussed prominently in May 2014 by Cisco's CEO John Chambers, who predicted "brutal, brutal" IT consolidation at his Cisco Live customer and partner event in San Francisco.[14] He said that the pace of change in IT is accelerating, and that we "will see this disruption not so much in consolidations, but almost like musical chairs." So in the HPC vendor community, China's NUTC, IBM, Cray, and Dell occupy the top-ten chairs of the TOP500, with the next top tier occupied by NEC, HP, SGI, Bull, Fujitsu, and Hitachi. Naturally, many more chairs are occupied at a deep level of intellectual property and component manufacturing.

Further disruption: on May 26, 2014, France's Atos, arguably an IT cloud and software company, announced its intent to purchase computer company Bull.[15] Founded in 1930, Bull once rivaled IBM in calculating machines. After decades of bailouts involving GE, Honeywell, Motorola, NEC, Orange, and the French government, its new chair will be occupied by a modern combined company that earns its greatest fraction of revenues from consulting and systems integration.

As the supercomputing community moves toward exascale (1000 times faster/larger than petascale), who will be the winners and losers? And to whom do the application vendors hook their wagon?

We shall see.

ADDITIONAL READING

1. Bell G. "The Future of High Performance Computers in Science and Engineering", Communications of the ACM, Vol 32, No. 9, September 1989.
2. Bell G. "Ultracomputers: A Teraflop Before Its Time", *Communications of the ACM*, Vol. 35, No. 8, August 1992.
3. Bell G. "1995 Observations on Supercomputing Alternatives: Did the MPP Bandwagon Lead to a Cul-de-Sac?", Communications of the ACM, Vol. 39, No. 3, March 1996.
4. Bell G. "A Brief History of Supercomputing: 'the Crays,' Clusters and Beowulfs, Centers. What Next," Bay Area Research Center, Microsoft Research, http://research.microsoft.com/en-us/um/people/gbell/supers/supercomputing-a_brief_history_1965_2002.htm.
5. Susan G, Marc S, Cynthia AP (Editors). *Getting Up To Speed: The Future of Supercomputing*, Committee on the Future of Supercomputing, Computer Science and Telecommunications Board, Division of Engineering and Physical Sciences, National Research Council of the National Academies, The National Academies Press, Washington, D.C., 2004.
6. Foster I, Kesselman C (Editors). *The GRID: Blueprint for a New Computing Infrastructure*, Morgan Kaufman, San Francisco, 1999.
7. NSF Press Report "Prospectus for Computational Physics" [Report by the Subcommittee on Computational Facilities for Theoretical Research to the Advisory Committee for Physics, Division of Physics, March 15, 1981].

8. Murray C. The Supermen: The Story of Seymour Cray and the Technical Wizards behind the Supercomputer, *Design News Magazine*, 1997.
9. Grier, David Alan, The Human Computer and the Birth of the Information Age, Joseph Henry Lecture, Philosophical Society of Washington, May 11, 2001.
10. Freiberger, Paul and Swaine, Michael, *Fire in the Valley: The Making of the Personal Computer*, McGraw-Hill, Second Edition 2000, ISBN 0-07-135892-7.
11. Redmond, Kent C.; Smith, Thomas M. (1980). *Project Whirlwind: The History of a Pioneer Computer*. Bedford, MA: Digital Press. ISBN 0-932376-09-6.

REFERENCES

1. Metropolis N, Howlett J, Gian-Carlo R (Editors). *A History of Computing in the Twentieth Century,* Academic Press, New York, NY, 1980, Chapter Title: Computer Developments 1935–1955, as Seen from Cambridge, USA, by Garrett Birkhoff. This book contains the edited versions of the papers presented at the International Research Conference on the History of Computing, held at the Los Alamos Scientific Laboratory, June 10–15, 1976.
2. Redmond KC, Thomas MS. From Whirlwind to MITRE: The R&D Story of The SAGE Air Defense Computer. Massachusetts Institute of Technology, Cambridge, Massachusetts, 2000.
3. Flamm K. *Creating the Computer: Government, Industry, and High Technology*, Brookings Institution Press, Washington, D.C., 1988.
4. Sterling T, Salmon J, Becker DJ, Savarese DV. *How to Build a Beowulf: A Guide to the Implementation and Application of PC Clusters*, MIT Press, Cambridge, MA, 1999.
5. Sterling T. *Beowulf PC Cluster Computing with Windows and Beowulf PC Cluster Computing with Linux*, MIT Press, Cambridge, MA, 2001.
6. Sterling T, Paul M, Paul HS. *Enabling Technologies for Petaflops Computing*, MIT Press, Cambridge, MA, July 1995.
7. A complete computer timeline can be viewed at the Computer History Museum site: www.computerhistory.org/timeline/
8. Matlis, Jan, Computerworld, *A Brief History of Supercomputers*, May 31, 2005, www.computerworld.com.au/article/132504/brief_history_supercomputers/
9. A detailed history of the NSF and the National Science Board can be found in the booklet *The National Science Board: A History in Highlights 1950–2000* and is available on the Internet at www.nsf.gov/nsb.
10. Black Proposal, *A Center for Scientific and Engineering Supercomputing*, 1982, http://www.ncsa.illinois.edu/20years/timeline/documents/blackproposal.pdf.
11. http://www.pnl.gov/scales/docs/lax_report1982.pdf.
12. http://www.nsf.gov/awardsearch/showAward?AWD_ID=8404556.
13. https://archive.org/details/federalsupercomp00unit
14. Prickett Morgan, Timothy, Enterprise Tech Networks Edition, *Cisco CEO Predicts 'Brutal, Brutal' IT Consolidation*, May 22, 2014.
15. Mawad, Marie and Rahn, Cornelius, Bloomberg.com, *Atos Bids for Bull to Restore Computer Maker to Past Glory*, May 27, 2014.

I

Country Contributions

CHAPTER 2

Country Strategic Overview of Supercomputing

Steve Conway,* Merle Giles, and Anwar Osseyran

CONTENTS

* IDC's HPC study for the European Commission, Steve Conway, IDC Research Vice President, High-Performance Computing/Data Analysis & Steering Committee Member, HPC User Forum (www.hpcuserforum.com).

2.1 EUROPE

In most Western European countries, high-performance computing (HPC) national programs obtain support for procurement and running of Tier-0, Tier-1, and Tier-2 HPC systems as either explicit funding programs or included in budgets of universities or research organizations. In addition to procurement and operations, countries vary when it comes to their specific strategic efforts and the extent of those efforts. As examples, Germany, France, United Kingdom, Finland, Norway, and Switzerland have each launched comprehensive and visible national HPC strategies, including substantial financial support for research in HPC, as described in Sections 2.2 through 2.7.

2.2 GERMANY

Germany's modern position in HPC was strengthened with the creation of the Gauss Centre for Supercomputing (GCS) that was the result of an alliance between three national Tier-0 centers—Forschung Zentrum Jülich (FZJ), the Leibniz-Rechenzentrum (LRZ) in Garching, and HLRS (Höchstleistungsrechenzentrum)—under a July 2006 agreement between the Federal Ministry for Education and Research (BMBF) and the Regional Ministries for Research of Bavaria, Baden-Württemberg, and Nordrhein-Westfalen. Approximately 50% of government investment in HPC in Germany comes from the German states (Länder), which fund half of the three Tier-0 centers. The computational science projects of the John von Neumann Institute for Computing (NIC) and the research activities of KONWIHR (Bavarian Competence Network for Technical and Scientific High-Performance Computing) bring together mathematicians, informatics, and application domains with emphasis on the interdisciplinary development of new application codes.

At the national level, a massive BMBF investment in the D-Grid project* (€100 million over 5 years) resulted in a core grid infrastructure for German scientific collaboration. Recently, strong emphasis has been dedicated to HPC software research projects within the German HPC Software Initiative under the "IKT 2020 Förderprogramm." The innovation policy of the German Federal Government is focused on technology developments aimed at scientific and technological leadership, and the integration of novel services. HPC is clearly seen as needed for these goals. The Call for "HPC Software for Scalable Parallel Computers"

* http://www.d-grid-gmbh.de/index.php?id=1&L=1.

in January 2010 was intended to close the gap between HPC hardware performance and application software scalability on extremely parallel systems, the major barrier to progress in computational science and engineering in simulation-based HPC applications.

As the complexity of extremely parallel systems and applications requires interdisciplinary cooperation among experts in the application fields, mathematical methodology and computer science, the invitation of the new BMBF to submit HPC software proposals requires the projects to act as active nodes on an HPC software network to cooperate within the existing Gauss Alliance of local and regional computer centers and the national supercomputing centers. In particular, effective cooperation between the HPC community and users of commercial applications is expected.

2.3 FRANCE

France has long invested in HPC and simulation, particularly in the defense sector. A new impetus was provided in 2005 with the creation of a strategic committee for HPC, the CSCI (le Comité Stratégique du Calcul Intensif), the creation of GENCI (Grand Equipement National de Calcul Intensif), which is entrusted with funding and ownership of major computer equipment for the French computer centers for civilian research, and the launch by ANR (Agence Nationale de la Recherche) of the call for projects, "Calcul Intensif and simulation," now called Program Cosinus. According to The French ANR, the total ANR funding for research in HPC is €25 million a year.

ANR published a book in January 2010 to raise awareness for HPC research in France—"Le calcul intensif: technologie clé pour le future." This work reported that 37 out of 123 projects were categorized as belonging to HPC-related "competitiveness clusters." Three competitiveness clusters—System@tic with nine projects, Materalia (innovative materials for intelligent products of the future, Lorraine and Champagne-Ardennes region), with five projects, and Aerospace Valley (Toulouse Region with four projects)—concentrated 50% of funding in terms of number of projects in 2009.

Competitiveness clusters—in particular System@tic—and large research labs—CEA (Commissariat à l'énergie atomique), INRIA (Institut national de recherche en informatique et en automatique), and CNRS (Centre national de la recherche scientifique)—are where most research in HPC is carried out in France. Competitiveness clusters are mainly funded by the

French government via the FUI (Fond unique interministériel), but also by local government and the European Union through FEDER (European Regional Development Fund). Created at the initiative of CEA-DAM Ile de France, TER@TEC association, a member of the System@tic Paris Region Competitiveness cluster, has brought together more than 60 companies and research labs and built a high-performance digital simulation competence center that has initiated or taken part in major HPC research projects within System@tic, such as POPS, Open HPC, or CDSL.

2.4 UNITED KINGDOM

The coordination of HPC activities for academic research in the United Kingdom is the joint responsibility of the U.K. Research Councils. Acting on behalf of the Office of Science and Technology (OST), the Engineering and Physical Sciences Research Council (EPSRC) implements and runs the U.K. research councils.

In 2007, the High End Computing Strategic Framework Working Group produced a Strategic Framework for High End Computing (HEC) at the request of the High End Computing Strategy Committee. The roadmap includes the strategic direction of HPC developments in the United Kingdom, access arrangements to HPC services, planning the procurement and location of HPC services, and promoting their widespread use.

In terms of HPC research, it was decided that new initiatives and funding should be established to grow the discipline of computational science and engineering (CS&E) within the United Kingdom as "an advanced and complex HEC e-Infrastructure is ineffective without the people with the expertise to utilize it." The HEC strategy also called for partnerships between U.K. organizations, including the Research Councils, industry and the Meteorological Office to produce economies of scale and to promote new opportunities for research and the technology transfer. The HEC strategy also encourages interdisciplinary coordination activities between HPC consortia that participate in U.K. Collaborative Computational Projects. These Collaborative computational projects bring together the major U.K. groups in a given field of computational research to tackle large-scale scientific software development projects, maintenance, distribution, training, and user support.

The HEC strategy resulted in the HEC program that is coordinated by the EPSRC. The two main centers supported are EPCC (Edinburgh Parallel Computing Center) that manages both U.K. national HPC facilities (ARCHER and DIRAC) and CLRC-Daresbury Laboratory. The Hartree Center is a flagship initiative. It is currently being implemented as a new kind of computational sciences institute for the United Kingdom. It will bring together academic, government, and industry communities to focus on multidisciplinary, multiscale, efficient and effective computation focused among others on the themes of energy, climate, health, and security.

2.5 FINLAND

A leading country in computing and simulation, Finland has long developed a strategy to foster both science and industry competitiveness with HPC. Its unique large national center, CSC, provides both scientists and industry with HPC resources and runs grid projects with Finnish universities that host much smaller centers. These universities collaborate in a well-recognized efficient way with industry in a wide array of scientific disciplines including HPC. The main public funding organization for research and development in Finland, Tekes (Finnish Funding Agency for Technology and Innovation), funds industrial projects as well as projects in research organizations, and especially promotes innovative, risk-intensive projects in scientific areas defined at the national level. Tekes allocates about half the financing granted to companies, universities, and research institutes through the programs.

The flagship program in Finland was its 5-year MASI (Modeling and Simulation) program.* The aim of MASI, launched in 2005, was to develop new modeling and simulation methods, to enhance the utilization of these technologies in industrial and service sectors, and to create a competitive edge for Finnish companies in global markets. The program catalyzed new businesses based on multidisciplinary modeling and simulation expertise. The total cost of MASI program research projects amounted to €100 million over 5 years, with Tekes' share amounting to €53 million. Other HPC or HPC-related projects are also funded by the Academy of Finland.

2.6 NORWAY

Norway has long been strategic on supercomputing with the Research Council of Norway (NFR) defining and implementing HPC policy. Launched in 2000, the Notur project provides the national infrastructure for computational science in the country to researchers at Norwegian

* http://www.csc.fi/english/csc/publications/cscnews/2010/2/masi.

universities and colleges, and operational forecasting and research at the country's Meteorological Institute. Half of the Notur project funding came from NFR and the other half from a consortium of public and private partners including major universities and industry. The NFR-funded program eVITA provides for research projects aimed at developing new theories, new models, methods, algorithms, techniques, and tools for applying high-volume computing and data resources to problems in science, technology, and medicine. Following through on the statement "The most scientifically important and economically promising research frontiers in the 21st century will be conquered by those most skilled with advanced computing technologies and computational science applications," sixty percent of the programs' research component is dedicated to large-scale interdisciplinary projects that benefit from an annual budget of NOK 110 million, or nearly $18 million.*

2.7 SWITZERLAND

Top HPC research has long been performed in Switzerland by its Federal Institute of Technology ETH Zurich (Eidgenössische Technische Hochschule Zürich) and EPFL (École polytechnique fédérale de Lausanne). In May 2009, the Bundesrat launched its HPCN (HPC+Networking) strategic plan positioning HPC high in the research and competitiveness agenda in Switzerland. HPCN was to be implemented over 3 years at a cost of €120 million. The third pillar of this strategy was research and education to enable scientists to efficiently use the future research infrastructure to be implemented within the framework of the strategy. This resulted in the creation of an HP2C Swiss Platform for High-Performance and High Productivity Computing† that ran until the summer of 2013. Part of this effort was dedicated to developing applications to run at scale and to make efficient use of the next generation of supercomputers. Its follow-on initiative is called Platform for Advanced Scientific Computing (PASC) and covers the period 2013–2016.

2.8 ITALY AND SPAIN

Italy and Spain have not launched national HPC planning strategies comparable to other European countries. However, they have funded and now host some of the largest HPC centers in Europe, including Cineca

* http://www.forskningsradet.no/prognett-evita/Programme_description/1226485583597.
† http://www.hp2c.ch/.

in Bologna, Italy* and Spain's Barcelona Supercomputing Center (BSC) established in 2005†, Centro de Supercomputació de Catalunya (CESCA), Centro de Supercomputación de Galicia (CESGA), and others. Research in HPC is substantial in both countries. BSC cooperates with worldwide leading HPC technology vendors and performs leading research in several HPC research domains. In Italy, research is disseminated in a larger number of research labs and institutes such as ICAR (Istituto di Calcolo e Reti ad Alte Prestazioni) and Cineca. These universities and centers are also active participants in EU-funded projects.

2.9 RUSSIA‡

HPC Wire's Tiffany Trader reported in December 2013 that the supercomputing arms race is in full force as nations awaken to HPC's potential. The countries listed above awakened early, but in 2009 Russian President Dmitry Medvedev called for rapid development of supercomputers in Russia. His speech to the Security Council was reported by news outlet ITAR-TASS and archived on the Kremlin's website: "We have to work at stimulating demand (for supercomputers) in every possible way, not because this is a fashionable topic, but simply because if we don't create such a demand our products will not be competitive or of interest to potential buyers." He continued: "Once again any sort of airframe or engine that is not produced with the aid of supercomputers is unlikely to trigger interest among buyers in a few years, because even now there are standards already set and so far we are doing practically nothing to meet them."

The speech triggered a wave of support for HPC in Russia, and less than 2 years later, the nation celebrated its first petascale system. An upgrade to Moscow State University's "Lomonosov" supercomputer increased its peak theoretical operating speed to 1.3 petaflops, scoring the T-Platforms system a 13th-place finish on the June 2011 TOP500 list and a no. 37 ranking in November 2013.

Sergei Abramov, head of the Russian Academy of Sciences Program Systems Institute in Pereslavl-Zalessky, an historic town 140 km north east of Moscow, went so far as to say "a supercomputer is the only instrument to beat a competitor" as he recalled words spoken by U.S. Council on Competitiveness' President Deborah Wince-Smith: "the country that wants to outcompete must outcompute."

---

* http://www.cineca.it/en.
† http://www.bsc.es/.
‡ http://www.hpcwire.com/2013/12/03/russia-aims-close-supercomputing-gap/.

HPC Wire's Trader goes on to report that the most difficult part for countries attempting to "outcompute" is developing their own "homegrown" supercomputing technologies. This capability provides the ultimate competitive advantage over other nations, according to Abramov. Currently, Russia must rely on commercially available components from other countries, most notably for its chips and storage, which Russia has not developed internally.

2.10 THE NETHERLANDS, SWEDEN, AND DENMARK

Northern European countries such as The Netherlands, Sweden, and Denmark have defined and implemented HPC strategies based on research infrastructures and grids to support their scientific communities. They have provided medium- or long-term funding for the installation, running costs, and future upgrades or replacements of their infrastructures. These countries carry out research efforts in HPC within universities' computational science departments, supercomputing centers, or public–private clusters, a high proportion of which is carried out within EU-funded projects.

2.11 BELGIUM, AUSTRIA, AND POLAND

In Belgium, research is undertaken more at the regional level, such as at Wallonia's Cenaero Center of Excellence in Aeronautical research* (with funding from the EU FEDER program) and Flanders' Vlaams Supercomputer Centrum (VSC) hosted at Ghent University. Two other countries should be mentioned for their substantial "research in HPC" activities within their universities and internationally: Austria and Poland.

2.12 GREECE, CYPRUS, AND TURKEY

Greece has been a long-time HPC participant through institutions such as the University of Athens, the National Technical University of Athens,† and the Computer Technology Institute of the University of Patras, as well as through participation in EU collaborations.

The Cyprus Institute plays a major role in supporting HPC initiatives and international collaborations, and boasts a talented, experienced HPC research community. Strategic partnerships have been formed with

* http://www.cenaero.be/.

† http://www.cslab.ntua.gr/.

Massachusetts Institute of Technology (MIT) and University of Illinois at Urbana-Champaign (UIUC), as well as the Centre de recherche et de restauration des musées de France (C2RMF) and Germany's Max Planck Society.*

Turkey inaugurated its National Center for High-Performance Computing in 2004 as a resource for both scientific and industrial research.†

2.13 JAPAN

The widespread popularity of The IBM System/360, introduced in 1964, impressed on the Japanese government the need to support its own computer industry, and in 1966, the Ministry of Economy, Trade and Industry (METI) launched a national project of "the development of the highly advanced computer". Hitachi, as a chief organizer, NEC Corporation, Fujitsu, Toshiba Corporation, Mitsubishi Electric Corporation, and Oki Electric Industry joined this project, which ran 5 years and consumed public funds amounting to around 10 billion yen, or roughly U.S. $100 million. Hitachi developed mainframe computers in 1970 and 1971 that were installed at major universities and national laboratories.

Japanese supercomputers reached their highest levels of performance in the mid-1990s, thanks to continuous research and development by supercomputer manufacturers and strong demand from the Japanese manufacturing industry. The Japanese economy was booming and many companies were trying to introduce Computer Aided Engineering (CAE) in their production processing as a capital investment to cut and shorten development costs and time. At the same time, however, the world-leading supercomputer technologies of Hitachi, NEC, and Fujitsu became a source of trade friction with the United States, and together with the subsequent collapse of the economic boom that resonated throughout the 1990s, resulted in some makers' withdrawal from the supercomputer business.

2.14 CHINA

China has been involved in developing supercomputers at least since 1983 and began using them even before then. China has boosted support for HPC as an economic accelerator and a symbol of technical sophistication.

* http://www.cyi.ac.cy/index.php.

† http://www.uybhm.itu.edu.tr/.

China's HPC market depends heavily on the dynamics between the central government and the governments of the largest cities and their surrounding regions. These parties often co-fund large supercomputer facilities and machines, with the cities sometimes contributing the larger share of funding and gaining access to the majority of the compute cycles. Big cities are the economic engines of the country and exert considerable power in these arrangements. They typically apply HPC to local/regional industries and to urban issues such as traffic management.

Most large HPC systems in China today use standard components from Intel, Nvidia, and so on, but there are at least three Chinese processor initiatives under way. China is now the second-largest country where HPC spending is concerned (United States remains no. 1). China has also been pushing HPC in the cloud hard, especially through government-funded cloud initiatives that make HPC resources available to science and industry.

2.15 SOUTH KOREA

South Korea has had two significant HPC sites: Korea Meteorological Administration and KiSTi. The Korean parliament mandated in 2010 that its government should invest more in HPC because of HPC's importance to science, industry, and national security. The country is now significantly expanding its HPC investments.

2.16 INDIA*

India's Center for Development of Advanced Computing (C-DAC) was set up in 1988 with the explicit purpose of demonstrating India's HPC capability following the denial of import of supercomputers from the United States. C-DAC has emerged as a third-party R&D organization in India's IT&E (Information Technologies and Electronics) working on strengthening national technological capabilities in the context of global developments in the field and responding to change in the market need in selected foundation areas.

2.17 AUSTRALIA†

HPC in Australia comprises an array of nine high-end computing facilities ranging from research petascale systems to modest university systems.

* http://cdac.in/index.aspx?id=hpc_grid.

† http://www.australianhpc.org.au/.

2.18 UNITED STATES OF AMERICA

American supercomputing providers can be segregated into two major branches: mission laboratories and open research. Laboratories with specific missions include those funded by the federal departments of defense, energy, and space. Defense-oriented mission labs at the Department of Defense (DOD) have secret, classified missions and will not be covered in this book. Likewise, the contributions of NASA (National Aeronautics and Space Administration) will not be explored here. There are a number of suitable books, both historical and technical, that cover these two agencies' accomplishments on the development of complex platforms.

Department of Energy (DOE) mission laboratories are managed by the NNSA (National Nuclear Security Administration) with responsibilities for nuclear management and security, such as weapons, energy, naval power, stockpiling, emergency response, and overall safety. DOE's open research is conducted in the ASCR (Advanced Scientific Computing Research) program, with responsibilities to discover, develop, and deploy computational and networking capabilities to analyze, simulate, and predict complex phenomena important to the DOE. Three specific NNSA labs, known as the TriLabs, are described in Chapter 14.

Other open research labs include supercomputer centers funded by the National Science Foundation (NSF) and universities. The reader will have a suitable understanding of the history from Chapter 1, with more detail available in Chapters 12 and 13. Together, the DOE, DOD, NASA, and NSF fund the largest, most powerful supercomputers in America.

Beyond the federal labs and NSF centers, there is a significant amount of supercomputing being conducted inside America's corporations. The extent is neither well understood nor well documented, for reasons that vary, but include at least the following:

- Industry work is often proprietary, and therefore not disclosed without explicit permission.
- Supercomputers in industry are funded privately and are similarly not disclosed without permission.
- Most companies do not choose to publicize the extent of their computing capacity for competitive reasons.

CHAPTER 3

Industrial Applications on Supercomputing in Finland

Per Öster, Anni Jakobsson, and Ari Turunen

CONTENTS

3.1 HISTORY OF HPC IN FINLAND STARTED IN SAUNA

The history of HPC in Finland began more than 40 years ago when the Bank of Finland had a surplus of 20 million Finnish marks in 1969. The bank asked the Finnish Innovation Fund Sitra if they could use the money for the public good. Professor Osmo A. Wio, while hosting a sauna evening, relayed the Bank of Finland's offer to his guests, whereafter professor Eino Tunkelo immediately suggested using the money to buy a supercomputer. This resulted in the purchase of a Univac 1108, which became operational in 1971. This was the start of the predecessor of CSC – IT Center for Science Ltd (Figure 3.1a and b).

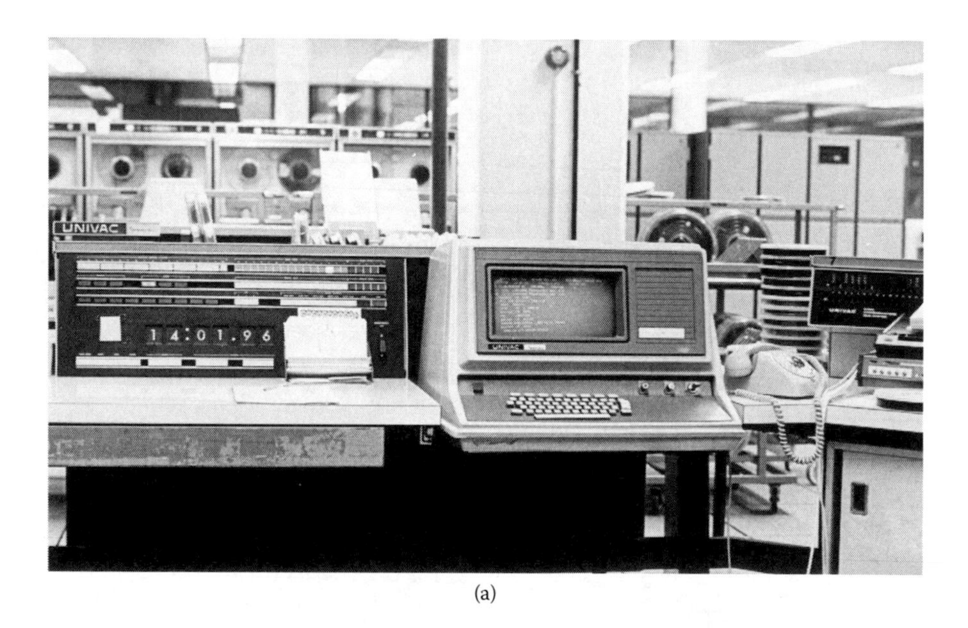

(a)

(b)

FIGURE 3.1 (a) and (b) Univac 1108 back in 1971 © CSC archives.

3.2 COMPUTER FACILITIES IN THE NORDIC COUNTRIES

In the early 1970s there were a couple of other computers at Nordic universities. In Sweden, the big central computer was located at the premises of the armed forces of the nation. The universities in Sweden did not have their own computers but used the central computer with a fast terminal connection.

It was also possible for the researchers in the Nordic countries to use the supercomputer—IBM 7090—located in Copenhagen at the Northern Europe University Computing Center for free. To use the machine, people had to travel to Copenhagen, as telecommunications networks did not exist. Professor Auvo Sarmanto remembers carrying magnetic tapes by plane to be loaded into the IBM 7090 computer as a courier from the University of Helsinki's Department of Nuclear Physics to Copenhagen.

3.3 HPC IN FINLAND STARTED WITH PAPER MACHINE ANALYSIS

From the birth in the 1970s to the present, HPC in Finland has been to a large extent an academic business. The first industrial HPC experiments in Finland were done in the 1980s in collaboration with universities and governmental R&D institutes. The first simulation was a vibration analysis of a full paper machine in 1986 for Valmet. For a long time, paper and pulp was the leading industry in Finland, and Valmet was one of the biggest companies in that field at the time.

The first industrial users generally performed linear structural analysis that is routinely done today with standard workstations or PCs. For example, Saab-Valmet performed an automotive vibration analysis of Saab Cabriolets. There was also offshore submarine development with Rauma-Repola Marine industries that at the time manufactured the Mir submarines that could descend 6000 meters. These submarines were later used in the filming of the movie *Titanic*.

With time, the need for supercomputers for structural mechanics has gradually decreased, with computational fluid mechanics becoming the main application field for industrial use. It was again Valmet that started to model critical parts of the paper machines in the mid-1990s. The focus was particularly on the optimal design of a paper machine headbox that is responsible for the even distribution of the pulp in the manufacturing process. There were also new industries that used supercomputing. In the early 1990s, Okmetic Ltd. started to model the single-crystalline silicon growth process. This is a complex process that includes a plethora of different physical phenomena. The problems were modeled in several stages for more than 15 years. Often, complex physical phenomena and supercomputing aspects cannot be separated. Sometimes the emphasis would be on sheer computing power but often the modeling itself was the bottleneck. The silicon industry's success also propelled the modeling of microelectromechanical systems (MEMS) at the start of the next millennium.

3.4 CURRENT PROVISIONING OF HPC IN FINLAND (THE NATIONAL HPC POLICY AND THE LOCAL ECOSYSTEM)

For more than 40 years, CSC has provided the Finnish research community with an open-access competitive Information and Communication Technology (ICT) infrastructure for research. With a continuous growth since its establishment in 1971, today the organization has over 250 employees and an annual turnover above €30M. CSC is a limited company owned by the Finnish Government and administered by the Ministry of Education and Culture. Finnish research trusts CSC through one or several of its services: the Finnish University and Research Network (Funet), supercomputer and computing clusters, storage and data services, software and tools, and training and expert support. The capabilities of CSC have enabled a number of advancements for Finnish research such as simulation of galactic dynamo processes, discovery of fundamental aspects of gold clusters and particles, understanding of aerosol effects on global warming, analysis of genetic variation in Finland and the Nordic countries, resolving 3D structures of viruses and more.

With a vision to "pioneer the sustainable development of ICT services" CSC has a further role to introduce new technologies and advances within ICT and its application to benefit Finnish research. This has been accomplished over the years by, for example, connecting Finland to the Internet in 1988, introducing the massively parallel Cray T3E computer in 1996, establishing optical research data network allowing link speeds up to 100 Gbps in 2009, building a modular free-cooling energy-efficient datacenter in 2012, and piloting a high-density supercomputer prototype with many-core accelerators in 2013. In 2014, the Cray XT40 system was upgraded to meet the latest technology standards. The second datacenter module was also completed in November 2014.

When building and providing a national trusted and dependable HPC infrastructure, good security is key. Central HPC resources, such as CSC's, that are to be accessed remotely need to be secure but still easy to access and use. Finnish R&D into data security is internationally recognized, with CSC and F-Secure performing pioneering work on security issues. Funet Computer Emergency Response Team (Funet CERT) of CSC was the first CERT in Finland, founded in 1995. The security services of F-Secure Oyj, founded in 1988, are used globally through more than 200 Internet service providers and mobile phone operators.

NIC.FUNET.FI or FTP.FUNET.FI, created especially for the Funet member community, is one of the world's largest and oldest archives for open information and software. It started in 1990, and in 1991 it became the "home" of the original Linux operating system. Its first web server was set up in 1992. The material available from the server included software from several different application areas, information about various sciences, Internet communication standards, newsgroup archives, and a lot of other information. Because it enabled low-cost management of capable hardware resources, the popularity of the Linux operating system has also clearly played a role in helping HPC spread more easily around the world. The CSC Funet archive is still in operation, providing researchers and students all around the globe with open-source software collections such as GNU/Linux distributions and databases. (GNU is a recursive acronym meaning "GNU's not Unix."*).

3.5 FINNISH HPC ECOSYSTEM (NATIONAL HPC CENTER: INFRASTRUCTURE AND SERVICES)

CSC's data center capability was expanded in 2012 with a new modern and modular data center situated in Kajaani in mid-eastern Finland, 500 km northeast of Helsinki. In November 2014 the second datacenter module was completed, to be taken into use in 2015. Data is one of the key resources for a modern day researcher. In the HPC datacenters, supercomputers use and produce data in their analyses and simulations and the storage systems are used to organize, keep and enable access to the data. The computing and storage systems need to be tightly connected for this to work smoothly and specialized high-performance networks ensure this. The new Kajaani datacenter is CSC's third. The two others, DC1 and DC2 are both located in Espoo. DC1 is a high-security and high

* https://gnu.org/. Retrieved 12/29/2014.

FIGURE 3.2 Sisu, the new Cray XC40 system located in the new CSC datacenter in Kajaani © Cray Inc.

availability data center primarily for storing, archiving and networking resources. DC2 is a standard versatile computer room with redundant power and cooling capacity.

The construction of a new cost-effective data center was driven by several factors: space, energy cost, and environmental concerns. The energy consumption of CSC during the last 6 years shows an average annual increase of 1.1 GWh (Gigawatt Hour). It would not be economically feasible to meet this demand by expanding the datacenter facility in the Espoo Otaniemi area. So, when expanding, a site was sought that could offer abundant, environmentally-friendly power source and space for future expansion. The Kajaani site is used to host a pulp and paper industry, so it has the benefit of local hydroelectric power production, plus its northern latitude makes "free-cooling" of the computer equipment possible. In an average year in Kajaani, 80%–85% of the time humidity and temperature is such that a temperature of 10°C can be achieved by evaporating water—a standard requirement for free cooling. The datacenter in Kajaani has been built using the latest technology in prefabricated datacenter modules. This creates a flexible and very power-efficient solution that can be expanded on demand with the lowest possible marginal cost (Figure 3.2).

3.6 INDUSTRIAL HPC APPLICATIONS LANDSCAPE (POTENTIAL INDUSTRIAL USERS: EARLY ADOPTERS AND PROMISING MARKET SEGMENTS)

The uptake of HPC by industry in Finland is naturally found within the basic industries like forestry and energy. An example of this is the CFD-HEADBOX project, a joint project between CSC, Valmet, Kemira, and Process Flow, funded by Tekes (The National Technology Agency of

Finland), which in the year 2000 was to implement computer modeling to optimize the shape of industrial instruments and their adjustment. The main focus of the project was on automatic optimization of industrial equipment and processes that engage flow systems. The aim of the project was to speed up the dimensioning of CFD-based fluid dynamics equipment by replacing the currently used trial-and-error approach with automated CFD optimization. The target is to integrate the optimizing program with the commercial fluid dynamics programs used by companies participating in the project. In Finnish industry, at least the dividing manifold in the paper machine headbox was dimensioned for each new instrument delivery using the CFD-based optimization.

3.7 PILOTS, LESSONS LEARNED, DOS AND DON'TS, AND SUCCESS STORIES

3.7.1 Elmer

Elmer is an open-source multiphysical simulation software mainly developed by CSC – IT Center for Science. Elmer development started in 1995 in collaboration with Finnish universities, research institutes, and industry. It is perhaps the most popular multiphysics simulation software published under open-source license (LGPL/GPL) and has thousands of users around the globe.

Elmer has been instrumental in bringing many of the industrial applications to the realm of supercomputing. Many of the industrial cases computed by CSC's supercomputers have used Elmer as the primary tool. Elmer is a highly modular simulation software that utilizes the finite element method to solve the partial differential equations. The modularity and high level of abstraction has allowed new fields to be adopted without major changes in the core library.

Among the initial applications of the Elmer project was the simulation of Czochralski growth process of silicon. This already included several physical models that were implemented into the code. Thereafter, Elmer development was targeted to the simulation of MEMS and microfluidics. An important recent field of application was acoustics, where Elmer was used to study the acoustic losses in cellular phones.

Quite recently, Elmer has been developed to allow efficient modeling of electromagnetic problems in industry. ABB Motors and Generators in Helsinki is testing Elmer for its simulation of electrical machines.

ABB is a leader in power and automation technologies that enable utility and industry customers to improve performance while lowering environmental impact. To be able to model rotating machines, the challenging combination of computational methods had to be implemented in parallel. For example, mortar finite elements are used to ensure continuity of solution in the rotating interface and divergence-free edge elements are used to optimally capture the true magnetic fields.

The experiences with Elmer emphasize the importance of software in getting the most out of supercomputer investments. Both components are often needed to solve the most difficult problems. Also, industry that is trying to achieve competitive advantage cannot rely entirely on existing commercial software. On the other hand, today's hardest problems are easily solved in 10 years by standard tools and computers. Therefore, the research software must continuously evolve to target new problems.

3.7.2 FinHPC

The FinHPC project, sponsored by Tekes, contributed to the optimization of nearly 30 scientific program codes and helped to create a benchmark set to evaluate supercomputers. The project finished its work at the end of 2008. The positive impact of the FinHPC project is based on the collaboration in the development of the scientific applications. In the first phase of the project, a survey of optimization needs was carried out. Then codes were chosen and contact between the scientists who owned the code and the FinHPC group was established. Codes were profiled, tested, and selected based on importance, need for optimization and improvement opportunity. FinHPC project and its results quickly drew a lot of attention nationally and internationally and, even before the end of the project, the FinHPC partners become involved in efforts to establish a similar large-scale international collaboration. These efforts have now resulted in EU-funded large-scale projects on scientific simulation such as EUFORIA and PRACE.

3.7.3 SOMA

SOMA is a versatile modeling environment for computational drug discovery and molecular modeling developed together by CSC and Finnish small- and medium-sized enterprises (SMEs). Tekes funded the development process. Collaboration with SMEs resulted in at least three pharmaceutical substance patent families that brought significant income to the Biotie and Hormos Medical companies. The software is distributed under the terms of GNU General Public License since 2007.

3.8 FUTURE FOR HPC IN FINLAND

Finland is one of the most research-intensive countries in the world: more than 1% of the gross domestic product is spent on R&D, and 1.55% of the nation's workforce are researchers. So, obviously, the need for HPC is constantly growing in both academia and industry. As a whole, the Finnish research area has benefitted from the centralized model, concentration of HPC resources, and building a necessary critical mass of HPC knowledge at CSC. Still CSC's support to industry R&D is limited but there is for sure a potential to be exploited. This limited support is due partly to the company's place in the governmental system. The Ministry of Education and Culture hosts the country's largest supercomputers via CSC. As expected, the primary objective of the Ministry of Education and Culture is to satisfy the high-performance computing needs of higher education and research, resulting in the industrial needs of HPC being somewhat secondary. Therefore, CSC's role in supercomputing in industrial applications is rather small compared to many other countries with a similar industrial profile as Finland.

For a relatively small economy like Finland, the use of supercomputing also depends quite heavily on the industrial profile. In classical computational engineering, structural mechanics rarely requires supercomputing. In as yet demanding areas like fluid dynamics, Finland is missing the big automotive and airplane manufacturers that typically invest heavily in supercomputing. However, there is a strong process industry that does not yet fully utilize the possibilities of simulation, probably because a lot of the application areas are rather "dirty" with a complicated mix of phenomena making the application of computing more difficult. Also the electromechanical industry with many challenging coupled problems has a potential that has only recently started to be exploited. The industrial profile of Finland is constantly changing and, during the last years, many data-intensive applications have started to appear.

For academic research, HPC is of increasing importance, and new areas such as life sciences and statistics, and its various applications to data analysis, for example, "big data," are making use of the supercomputing resources. So, the future looks bright for HPC in Finland. The use within academic research is ever growing and the adoption in industry is good, but there are certainly opportunities to exploit. European collaboration is crucial for the future development of HPC on a national level. Finnish researchers have benefitted greatly from the Partnership for Advanced Computing in Europe (PRACE) where they have managed to get a proportionally large part of the available computing time on Europe's largest supercomputers. Finland has pushed for a strong education and training program within PRACE recognizing that giving the young PhD students strong HPC skills is to build for the future.

CHAPTER 4

Industrial Applications on Supercomputing in France

Catherine Rivière and Stéphane Requena

CONTENTS

4.1 HISTORY

In France, industry began to significantly use HPC in the early 1980s. Back then, the usage was mainly driven by big industrial companies: oil and gas companies such as TOTAL, Elf Aquitaine, and CGG were using vector machines for improving seismic processing, energy companies like EDF used HPC to model the fluid dynamics involved inside nuclear reactors, and car companies such as Renault and PSA were using it to perform CFD and crash test simulations. Similar adoption happened on the independent software vendors' side. It was the French group ESI that realized, in 1985, the first numerical crash test simulation ever done, for a consortium led by Volkswagen, and it was Dassault Systèmes, created in 1981, that pioneered computer aided design (CAD) with the tool that became a worldwide leader in this market: CATIA (originally named CATI).

In the 1990s, the use of HPC by industry began to conquer new domains, most notably aeronautics with Airbus and Dassault Aviation and finance with BNP Paribas and Société Générale.

Nowadays, although the historical domains mentioned firmly remain heavy users of HPC, the usage has spread across various industrial fields, and an increased number of industries are jumping on the bandwagon. Electronics, telecommunications, health, cosmetics, multimedia, and environmental companies, to name a few, are beginning to routinely rely on HPC to increase their productivity and cut costs. Aeronautics and car manufacturers use increasingly complex virtual prototypes to save millions of euros in physical prototypes, whereas at the other end of the spectrum, small- and medium-sized enterprises (SMEs) are seizing the opportunity by applying HPC to specific innovative applications like marine hydrodynamics for renewable energies.

The democratization of the use of HPC in industry is on good track in France, especially for big companies. However, there is still a lot to do with regard to expanding its use especially in the whole supply chain of those big companies that are now outsourcing the development of their products. As the same constraints, methodologies and tools are propagated to the full supply chain, the global competitiveness could be increased by fostering the use of advanced numerical simulation in all stages including SMEs.

4.2 CURRENT HPC SERVICE PROVIDER LANDSCAPE

The French HPC service provider landscape is very rich. It goes from providers from industry to providers from academia whose mission includes the objective to democratize the use of numerical simulation and HPC in industry.

4.2.1 HPC Service Providers from Industry

The development of HPC service offered to industrial companies is linked with the surge of interest in "cloud" applications. Companies, especially

SMEs that do not use HPC regularly enough or cannot invest in a system of their own, are turning to service providers to rent computing nodes for a specific period or even to offer their customers' products under the Software-as-a-Service model.

In France, there are two main industrial service providers. BULL is the main service provider. The company, created 80 years ago and very well-known in the HPC market for its HPC systems, is providing cloud and hosted HPC services to industry through a specific HPC offer called Extreme Factory.* BULL focused its offer around customizable services and technical support open all day long. BULL also offers a web portal that gives companies access to externalized computing resources, along with simulation and visualization applications. Integration with software vendors to propose the widest set of applications possible is key in this competitive market, and BULL has recently signed multiple partnerships with software vendors such as Ansys, Altair, CD-adapco, Distene, and ESI.

It is noteworthy that BULL is also, along with more than 44 partners, involved in the Fortissimo European project (http://www.fortissimo-project.eu), funded by the European Commission, aiming to set up a one-stop-shop for commercial HPC cloud services based on existing national initiatives.

The other service provider from industry is OVH, one of the leading providers of cloud services in Europe. OVH is now venturing into the HPC sphere, as demonstrated by the acquisition in October 2012 of Oxalya, a French company specialized in HPC. Through Oxalya, OVH, like BULL, is now proposing an offer for HPC services for industry (https://www.ovh.com/fr/hpc) spanning from on-demand cloud services to full hosting services.

4.2.2 HPC Service Providers from Public Research

4.2.2.1 At the National Level

The French HPC landscape in public research has greatly evolved recently: a single entity is now in charge of leading the French HPC strategy for research and of coordinating its major equipment, made available to French scientists for civilian research. Created in 2007, GENCI† is a civil society whose associates are the French Ministry of Higher Education and Research, research organizations CEA, CNRS, and Inria, and the Universities through the CPU (Conférence des Présidents d'Universités).

GENCI currently owns HPC systems operated by three national computing centers (TGCC from CEA, CINES from French Universities,

* http://www.extremefactory.fr.

† http://www.genci.fr/.

and IDRIS from CNRS) and these acquisitions have massively increased the national HPC resources, from just 20 teraflops in 2007 to more than 5.7 petaflops now, an 280-fold expansion beneficial to the community of French scientists.

GENCI's facilities are accessible to French scientists through a unified peer-review process based on scientific excellence. GENCI, whose mission includes promoting the use of numerical simulation and HPC both in fundamental and industrial research, makes it possible for industrial users to access its systems, mainly in the following two cases:

- For Open R&D projects: this mechanism helps to foster technological transfer between academia and industry. The industrial user must cooperate with an academic partner in a joint open research project. The academic partner is the principal investigator (PI) of the proposal, which is peer-reviewed using scientific excellence criteria. If their project is selected, the partners can use the systems for free, under the condition that they publish their results at the end of the 1-year grant period.
- For a "one-shot" experiment: to help industrial users to assess and demonstrate the potential of HPC in their daily business, or to perform unprecedented "Grand Challenges," industrial users may also use GENCI's HPC facilities.

This "one shot" possibility has been used by several large French companies including EDF, ArcelorMittal, and Veolia Environnement to name a few. Nonetheless, SMEs deserve a tailored integrated offer (not just raw access to HPC systems) and that is why, at the end of 2010, GENCI, Inria, and BPIFrance (a French public bank for developing innovation in industry and especially toward SMEs) decided to launch a specific initiative called "HPC-PME" (see Section 4.7).

The other main service provider for public research at national level is the Centre de Calcul Recherche et Technologie (CCRT),* a direct partnership between CEA and big industrial companies.

CCRT has been set up in 2003 by CEA to provide HPC resources for large scientific computations and to foster a real synergy between academia, other research organizations, and industry by promoting exchanges and scientific collaborations between all partners.

* http://www-ccrt.cea.fr/.

The current CCRT industrial partners are EDF (energy provider), SAFRAN (Aeronautics Company), EADS (Space Industrial), Areva (Global Nuclear Industry Leader), VALEO (Automotive Industrial), L'Oreal (Cosmetics Industrial), Thales Group (Aerospace, Defense, Transportation, Security) and INERIS (Industrial Risk Assessment Agency).

Each industrial partner signs a multiyear partnership contract whose duration is important to set up a fruitful collaboration between industrial users, the computing center team and other academic users in the same scientific field. In particular, the partners can share experiment outcomes, including experience of open-source or academic software.

Current HPC facilities of CCRT include a BULL ×86 cluster of 420 teraflops peak performance installed and operated at TGCC (in the south of Paris).

4.2.2.2 At the Regional and European Level

Similarly to the two main industrial offers from GENCI and CCRT that have been developed in the last ten years, there is a growing number of regional computing centers offering services to industries. And while the national level might be better suited for big companies or SMEs with already advanced knowledge in the field, the regional centers, with their proximity and their expertise, sufficiently satisfy the needs of local SMEs getting interested in HPC.

In 2011, GENCI conveniently started coordinating a project called Equip@Meso,* from the Investissements d'Avenir (literally, "investments of the future") program, that gathers 10 regional computing centers. Equip@meso is helping to structure the French HPC pyramid by providing funding for equipping the regional centers (called Tiers-2) in HPC facilities, by developing and sharing best practices in training and user support and by locally relaying the HPC-PME national initiative. Among these regional centers, some of them have significant experience in supporting HPC commercial services for industry. There are for instance CALMIP in Toulouse, ROMEO in Reims, CRIHAN in Rouen, or CIMENT in Grenoble.

GENCI acts as the French representative into the European research infrastructure PRACE (Partnership for Advanced Computing in Europe)† and is one of four hosting members together with Germany, Italy, and Spain. Since 2010, the supercomputer Curie has been made available to European

* http://equipameso.genci.fr.

† http://www.prace-ri.eu.

scientists by GENCI. Curie is a 2.0 petaflop BULL system operated by CEA and located at TGCC. In 2012, PRACE launched a specific offer toward industrial users called Open R&D. This offer allows them to use PRACE HPC resources and services for free, for the sole purpose of open research with publication of the results. Open R&D has already attracted more than 40 companies (from large companies to SMEs). Two years after the inception of the Open R&D offer by PRACE, France represents the biggest contributor in terms of number of companies using PRACE resources.

Similar to GENCI, a specific initiative toward European SMEs called SHAPE (SME HPC Adoption Program in Europe) has been established by PRACE on top of national initiatives like in France, United Kingdom, Germany, Italy, or The Netherlands.

Besides the SHAPE program, which will be further detailed in the SMEs section (Section 4.7), the FP7-funded Fortissimo* project has just started. Fortissimo aims to deploy a pan-European one-stop-shop HPC commercial cloud offering. This project started in July 2013 with 44 partners; among them, GENCI and INRIA are representing the HPC-PME initiative while BULL is also a major partner to propose at the French level an end-to-end solution spanning from evangelization to the use of commercial cloud services. It is coordinated by the University of Edinburgh with a budget of €16 million, mainly dedicated to performing, through call for proposals, up to 50 experiments of "cloudification" of industrial applications using Fortissimo services.

4.3 FRENCH HPC ECOSYSTEM

The French HPC ecosystem is organized around one top-down industry-oriented association, TER@TEC, and multiple bottom-up research-oriented associations that cover different aspects of the HPC field.

CEA, and particularly its military applications division (CEA-DAM), has long been a pioneering organization leading the democratization of the usage of HPC in France. The unique expertise that was developed through an intensive use of supercomputers, especially since 1996 when numerical simulations replaced real tests in the nuclear deterrence policy, was bound to benefit the whole of the French HPC ecosystem.

In 2004, CEA-DAM was the driving force behind the creation of TER@TEC,† an initiative now gathering 70 stakeholders from industry and

* http://www.fortissimo-project.eu.
† http://www.teratec.eu/gb/index.html.

academia spread along the whole value chain of HPC (from technology providers, independant software vendors [ISVs], and research labs to end users). TER@TEC is in charge of the promotion of HPC in industry and well known for organizing a very attractive yearly event with 400+ attendees in Paris in June. With the cooperation of local councils, TER@ TEC association is also setting up a technology park dedicated to high-performance simulation and computing, at Bruyères-le-Châtel, next to CEA's TGCC supercomputing facilities.

Other French organizations also play an important role in structuring the ecosystem:

- Groupe Calcul* is a service unit from CNRS in charge of training and dissemination of HPC in research but is more and more involved also with industry. Groupe Calcul also hosts the biggest French HPC mailing list registering 1200+ people from academia and industry. This mailing list is used for dissemination of conferences, job proposals, HPC announcements, ANR, PRACE, and GENCI calls for proposals and problems, hints and solutions provided by researchers.
- ORAP (Organisation de Recherche Associative en Parallélisme)† is an association created 20 years ago by CNRS, CEA, and INRIA in charge of dissemination of HPC in both academia and industry. ORAP is known in France for organizing a forum with 150+ attendees twice a year dedicated to promoting HPC technology, infrastructure, and applications for both research and industry.
- Finally, Aristote‡ is an association in charge of promotion of IT in France and organized across multiple working groups. The main focus of Aristote is the organization of seminars and trainings on topics that have a strong and innovative domain of interest for both research and industry. For HPC, this includes recent trainings on archive policies, Big Data, numerical scientific libraries, and collaborative visualization.

4.4 FRENCH HPC INDUSTRY

The French HPC industry is quite rich and encompasses both hardware and software HPC providers.

* http://calcul.math.cnrs.fr/.
† http://www.irisa.fr/orap.
‡ http://www.association-aristote.fr.

By far the best known is BULL, a worldwide leader in HPC solutions with major customers in France (GENCI, CEA, Météo France, Dassault Aviation, PSA, etc.), The Netherlands (SURFsara), Finland (CSC), United Kingdom (AWE), Germany (University of Aachen and Juelich Supercomputing Center), Spain (BSC), and Japan (IFERC).

Although BULL was almost nonexistent on the HPC market in 2005, it has rapidly grown into a major competitor and challenges the American vendors' quasi-monopoly. Its strength lies especially in its expertise on system-level software; BULL was one of the first providers to replace air-cooled systems by directly liquid-cooled systems, thereby greatly increasing the energy efficiency. Its success is also a result of multiple fruitful partnerships, especially with CEA, that allowed the company to gain valuable competences in system and architecture design. BULL also created sustained links with Intel and IBM to benefit from key electronic components like processors. BULL's added value is as much in the design of systems as in the tailored installation and maintenance of world-class supercomputers.

BULL is also providing remote HPC cloud services through an offer called Extreme Factory (see Section 4.2.1).

On the hardware side, it is also interesting to present KALRAY, a start-up founded in 2008, located in Grenoble and specialized in low-power manycore processors (especially the MPPA-256, a 5-watt 256-core processor) for specific markets in image and signal processing, telecommunication, intensive computing, industrial automation, data management and networking, and transport.

Besides hardware, France has always been a nurturing ground for successful software companies. The two main ones are Dassault Systèmes and ESI Group.

Dassault Systèmes, created in 1981 with presently close to 10,000 employees and revenue topping two billion euros in 2012, produces widely used software like CATIA, SolidWorks, ENOVIA, and Abaqus. It represents more than 20% of the lucrative CAD/CAE market.

ESI Group, created in 1973, is one of the leaders in the virtual prototyping market and its main clients include car companies (Renault, Ford, etc.), aeronautics (NASA, EADS, VZLU, etc.), and energy companies (Areva, EDF, etc.).

In addition to those two companies, it is worth mentioning smaller companies like Distene (specialized in preprocessing and postprocessing tools) and SAMTECH (now owned by LMS, developing a finite element base solver used for structural mechanics).

CAPS Entreprise, a spinoff of Inria, is internationally known for its expertise in programming models and tools for manycore and heterogeneous systems; two other spinoffs from Inria, Sysfera and Activeon, specialize in providing software solutions for application portals used in cloud and grid services.

4.5 INDUSTRIAL HPC APPLICATIONS LANDSCAPE

France is one of the countries in which the HPC capacity for industry is at the same level as the capacity of the research and academia sector, which usually has the lion's share.

Many large French companies are now using HPC as a tool to bolster their innovation and productivity, and some of them are among the world's largest users of HPC such as TOTAL, EDF, Airbus, CGG Veritas (not ranked the TOP500), Safran, Dassault Aviation, Renault, PSA, BNP Paribas, Société Générale, to name a few. The latest edition of the famous benchmark TOP500 (Figure 4.1), released in November 2013, demonstrates this: the French Company TOTAL owns the biggest HPC configuration (2.3 petaflops peak) in industry. Because new oil and gas reserves, located in complex geological zones are more and more difficult to discover, HPC has become a mandatory tool for increasing the success rate of wells drilled. Because the current average cost of a well drilled is $80+ million and exploring or exploiting an oil field requires multiple wells, the importance of using HPC for improving seismic exploration is unquestionable.

A recent IDC pilot study on the financial return on investment from the use of HPC showed that for the oil and gas industry, each U.S. dollar invested into HPC generates almost 357 dollars of revenue and 86 dollars of profit.

By consequence, this trend is not going to stop anytime soon, and TOTAL, like most of the big oil and gas companies (BP, ENI, and Petrobras), plans for major upgrades of its system in the years to come.

Aeronautics, with Airbus Group at its forefront, is one of the main sectors of usage in HPC. The engineering department of Airbus has been using HPC starting as far back as 1994, and its usage has steadily grown to tackle increasingly complex problems. HPC is a vital competitive advantage for the major companies; it leads to fewer costly prototypes and reduces time-to-market. The demand for HPC in the sector has moved from only CFD to include a whole range of problems such as the analysis of electromagnetic interferences to place antennas or resistance of composite

RANK	SITE	SYSTEM	CORES	RMAX (TFLOP/S)	RPEAK (TFLOP/S)	POWER (KW)
20	Total Exploration Production France	**Pangea** - SGI ICE X, Xeon E5-2670 8C 2.600GHz, Infiniband FDR SGI	110,400	2,098.1	2,296.3	2,118
73	EDF R&D France	**Zumbrota** - BlueGene/Q, Power BQC 16C 1.60GHz, Custom IBM	65,536	715.6	838.9	328.8
95	Airbus France	**HPC4** - HP POD - Cluster Platform BL460c, Intel Xeon E5-2697v2 12C 2.7GHz, Infiniband FDR Hewlett-Packard	34,560	516.9	746.5	
123	EDF R&D France	**PORTHOS** - IBM NeXtScale nx360M5, Xeon E5-2697v3 14C 2.6GHz, Infiniband FDR IBM	14,448	406.4	601.0	
142	EDF R&D France	**Athos** - iDataPlex DX360M4, Intel Xeon E5-2697v2 12C 2.700GHz, Infiniband FDR14 IBM	18,144	352.7	391.9	347.3
237	Airbus France	HP POD - Cluster Platform 3000 BL260c G6, X5675 3.06 GHz, Infiniband Hewlett-Packard	24,192	243.9	296.1	
338	Dassault Aviation France	BlueGene/Q, Power BQC 16C 1.6GHz, Custom Interconnect IBM	16,384	189.0	209.7	82
394	EDF R&D France	**Ivanhoe** - iDataPlex, Xeon X56xx 6C 2.93 GHz, Infiniband IBM	16,320	168.8	191.3	510
410	Financial Institution (P) France	iDataPlex DX360M4, Xeon E5-2670 8C 2.6GHz, Infiniband FDR IBM	8,864	167.7	184.4	184.1
411	Financial Institution (P) France	iDataPlex DX360M4, Xeon E5-2670 8C 2.6GHz, Infiniband FDR IBM	8,864	167.7	184.4	184.4
412	Financial Institution (P) France	iDataPlex DX360M4, Xeon E5-2670 8C 2.6GHz, Infiniband FDR IBM	8,864	167.7	184.4	184.1
424	Financial Institution (P) France	iDataPlex DX360M4, Intel Xeon E5-2680v2 10C 2.8GHz, Infiniband FDR IBM	9,520	163.9	213.2	175.2
476	Financial Institution (P) France	iDataPlex DX360M4, Intel Xeon E5-2680v2 10C 2.8GHz, Infiniband FDR IBM	9,180	158.1	205.6	192.8

FIGURE 4.1 TOP500 sublist (November 2014 edition) of HPC systems owned by industries in France.

materials to reduce the weight of the aircrafts. New constraints such as energy efficiency, greenhouse gas reduction, and noise reduction mean new applications for HPC.

In the framework of the FP7-funded EESI project,* Airbus and other companies (including TOTAL) made their HPC roadmap available toward exascale (Figures 4.2 and 4.3). It can be seen as a path to a fully virtual aircraft. As stated in this report, experts in this domain go even further: exaflops (10^{18} floating point operations per second), on course to be reached by the biggest systems in 2020, is not the ultimate goal and the timing must be in concordance with zetaflops (10^{21} floating point operations per second) systems close to 2030.

The aerospace industry is very lively on the HPC front as well. For instance, the demand in satellite launching has been boosted by mobile

* http://www.eesi-project.eu.

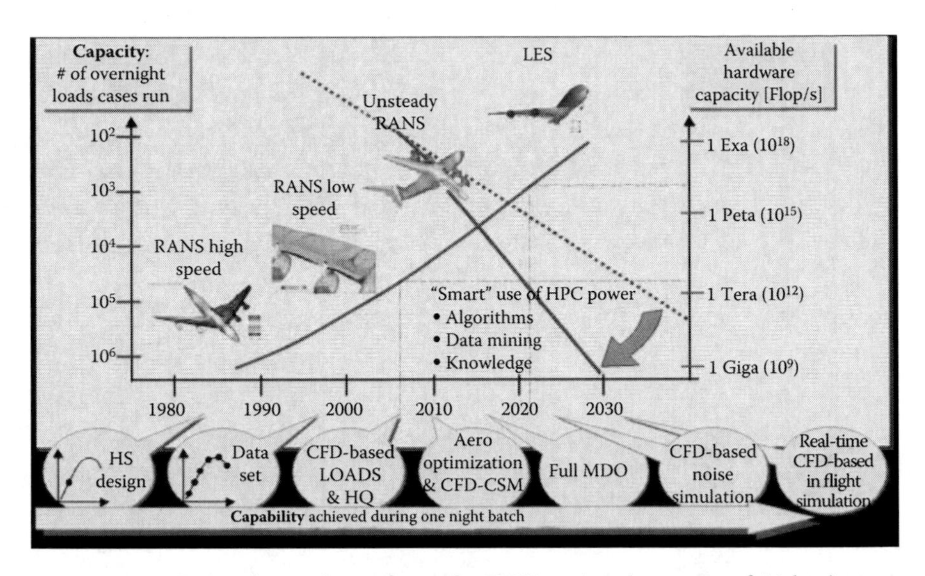

FIGURE 4.2 Exascale roadmap from the EESI project (courtesy of Airbus).

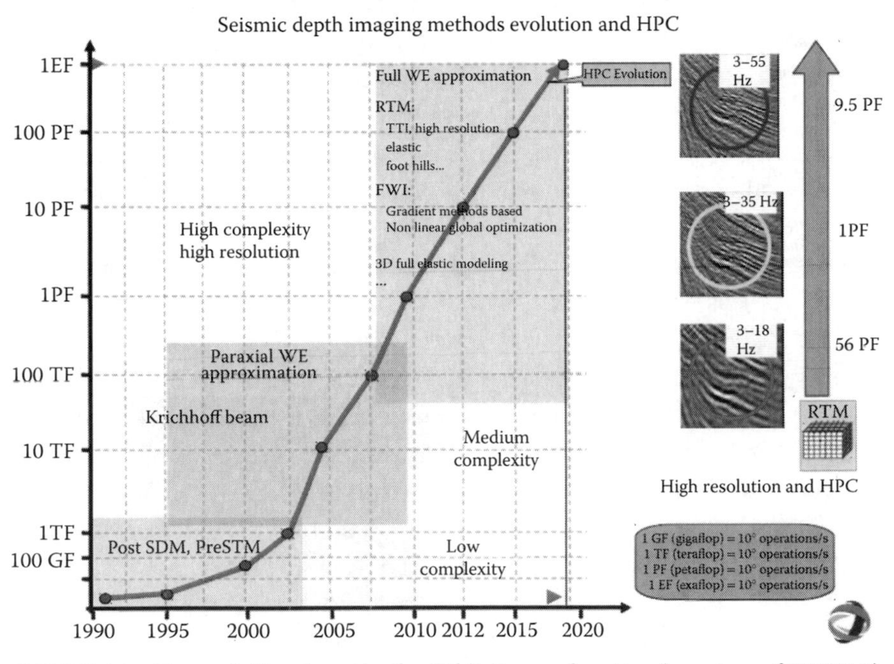

FIGURE 4.3 Exascale Roadmap in the Oil & Gas exploration (courtesy of TOTAL)

applications, high-speed internet, and global positioning needs, thereby driving up the software vendors operating on this market.

Most of what has been said for aeronautics also goes for the automotive sector: HPC is a vital innovation tool, to design better (more energy efficient, lightweight, etc.) and safer (crash tests) vehicles more quickly. Although some of the production has been shifted overseas to better answer the needs of developing countries, most of the R&D has stayed in France. HPC, once used only by the big car manufacturers like PSA or Renault, is spreading along the value chain. Big equipment companies such as Valeo, Faurecia, and PlasticOmnium also perform important HPC simulations and it is moving toward SMEs as well.

Energy is also a very important and historic sector for HPC. EDF, which currently owns the 46th and 257th biggest supercomputers in the world, uses HPC for its nuclear power plants (infrastructure design, safety analysis, CFD, energy optimization, process monitoring, material properties, etc.) but also for its hydroelectric facilities, its production of renewable energy, and other new subjects like Smart Grids. New environmental constraints to better assess the pollution and its possible impact have only reinforced the need for HPC. Many innovative French SMEs are also developing software for the production of renewable energy (for example Aria Technologies or OpenOceans for wind farm simulations) and energy efficiency (Izuba Energies for predicting energetic performance).

Other fields have now come to use HPC: banks and financial companies, especially for risk management and high-frequency trading, use HPC at least as much in their "production" environment as for R&D purposes; pharmaceutical companies like Sanofi use it to greatly improve their drug design by predicting the efficiency of thousands of potential molecules; cosmetics companies like L'Oreal, are integrating it in their product design—especially because the European Commission has recently forbidden animal testing. In a report issued by CCRT for its 10th anniversary, the L'Oreal representative in CCRT puts it this way: "For L'Oreal R&D, at the crossing of multiple disciplines, numerical simulation with HPC is becoming a key ingredient in the success of our main missions: discovering new grounds for innovation, predicting the efficiency and the safety of products, simulating to anticipate a significant performance."

4.6 CASE STUDIES AND BUSINESS CASES

HPC consistently improves companies' ability to shorten the time-to-market of new products, in many cases drastically lowering cost, improving the quality/reliability, and helping to explore uncharted business and technical territories. As shown by an IDC study, 97% of the companies that use HPC consider it indispensable for their ability to innovate, compete, and survive. Following are some examples illustrating the kind of benefits HPC can bring.

In the EESI project, Snecma, from Safran Group, stated that HPC has reduced the conception cycle of its helicopter engines by 6 months. Airbus has saved more than 20% of its costs in wind tunnel experiments thanks to numerical simulations. Each time TOTAL avoids a useless drilling to find oil, thanks to its analysis performed with HPC, it saves around €80 million. The Kaombo project that TOTAL ran in 2013 in Angola has produced 150 terabytes of data whose analysis now takes nine days with Pangea instead of 4½ months with the previous system.

Another example: Astrium (from Airbus Group) has developed a simulation software for the space launcher Ariane 5 that reduces the time needed to calculate the best trajectory for putting a satellite into orbit while optimizing the fuel consumption by the factor 10.

Some companies are also using the national and European research infrastructure to undertake Open R&D studies. In 2013, Renault was awarded 42 million core-hours on the French PRACE system Curie for an unprecedented multiphysics optimization of a car, using 10 times more parameters and applied to meshes 5 times bigger than the current state of the art. This study would have been impossible with their internal HPC facilities and thanks to these results Renault will be able to anticipate the future EuroNCAP6 regulations to come in 2016 and produce safer cars. For these results, Renault received the HPCWire Reader's Award for the Best Use of HPC in the Automotive Industry in November 2014.

In just a few years, the marine hydrodynamics SME HydrOcean, a spinoff of École Centrale de Nantes, went from first computations on regional computing centers to receiving the IDC "HPC Innovation Excellence Award" at SuperComputing 2013, the biggest annual event in the community. Guided by HPC-PME, it quickly climbed the ladder of the HPC ecosystem, and their projects were awarded with 5 million core-hours on a PRACE German system Hermit, and by 8 million core-hours on Curie. With 22 employees, HydrOcean is now visible at the worldwide level, awarded with new contracts with big automotive and naval companies, planning to hire 10 more people in 2014 alone and to open offices in Asia and South America.

4.7 STIMULATION PLANS FOR ADOPTION OF HPC BY THE FRENCH INDUSTRY AND SMEs

Although some of the big companies in Europe have already adopted HPC methodologies in their production cycle, SMEs, which represent the most important and dynamic part of the European Industry, are often either unaware of the potential of HPC for increasing their competitiveness or lacking expertise to know and apply new technologies for innovation such as HPC.

On the basis of this assessment, in 2010 GENCI, Inria (the French public research institute dedicated to computational sciences) and BPIfrance (a French public bank for developing innovation in industry and especially SMEs) conjointly launched the "Initiative HPC-PME" (HPC for SMEs Initiative) to allow French SMEs to concretely assess and demonstrate the potential of using advanced numerical simulation and HPC in their innovation process.

This initiative is supported by the five biggest technological clusters in France: Aerospace Valley (aeronautics and aerospace), Axelera (chemicals and materials), Minalogic (microelectronics), Systematic (embedded systems), and Cap Digital (digital media).

HPC-PME is based on an integrated offer providing high-value services like information, training, best practices, coaching and expertise from public research (in various scientific domains as well as in HPC), access to HPC national or regional centers, and funding facilities.

One of the most important added-values of this initiative is to foster technological transfer with academia thanks to the expertise provided by French research institutes such as CNRS, IFPEN, Ecole Potytechnique, and ONERA. HPC-PME connects SMEs with domain science or HPC experts from public research who will be able to understand the needs of the SME and then, using HPC, collaborate on an industrial project with the company. The proof of concept will provide the SME with a clear view on the return on investment to be expected when using these new technologies.

Since 2010, more than 50 SMEs have been followed within the framework of HPC-PME, from various sizes (2–500+ people) and various domains: automotive and aeronautics, oil and gas, offshore and sailing industry, finance, digital media and healthcare, microelectronics, and so on.

HPC-PME is now also part of the SHAPE (SME HPC Adoption Program in Europe) pilot launched by PRACE to help European SMEs to know and better use numerical simulation and HPC to increase their competitiveness. On the basis of the principles of complementary and eligible funding, SHAPE will rely on national initiatives when available (like in France with HPC-PME) or will directly support SMEs when not.

French SMEs enrolled in HPC-PME will be able, if needed, to scale out their industrial project on PRACE Tier-0 systems. It will also increase their visibility across a pan-European research infrastructure of 25 member states.

Another European project, Fortissimo, will cover the commercial part by developing a one-stop-shop HPC cloud offer. Ideally, if an SME has understood the benefit of HPC on its business through HPC-PME or SHAPE, and if it decides not to buy its own computing cluster, it can turn to Fortissimo to run simulations or offer services to customers on an HPC cloud infrastructure.

To further illustrate that governments are more willing than ever to encourage companies to adopt HPC, supercomputers were one of the 34 strategic plans outlined by the French Ministry of Industry in what it calls "the new industrial France." The development of this plan in the years to come is very likely to continue to support industries and especially SMEs in that direction.

4.8 EXTENSION TO BIG DATA

HPC appears to be the cornerstone technology to exploit the huge potential of big data. There cannot be any valuable analysis of large sets of unstructured data without a great deal of computing power and system expertise to empower it. Big data is a unique opportunity for the providers of service, hardware, and software to expand their business into new companies that have been well out of the market until now. Traditional users of HPC will also find new applications for their systems. For example, Orange, a French worldwide leader in telecommunications, explained that it was investigating how big data can enable them to anticipate customers' needs and problems.

In the oil and gas sector, as well as the automotive and aeronautics, the use of large-scale multiscale multiphysics simulation as well as the need to address uncertainties will generate a huge volume of data. New methodologies must be developed to analyze and store such pertinent data.

In life sciences and especially in genomics, the rise of next-generation sequencers, able to sequence a genome in just a few hours for less than $1000, is responsible for a data deluge that deserves specific data analysis and requires HPC and Big Data facilities.

The development of big data, in France as well as in other countries, will no doubt play an important role in shaping the future landscape of HPC.

4.9 CONCLUSION

As demonstrated by all the examples detailed in this chapter, and by the broad variety of industrial domains and types of companies now engaged in HPC, France and its companies are deeply committed to make this country one of the best in the domain, whether it be from a hardware, software, or application perspective.

Building on the early adopters in the 1980s and 1990s, and on the excellence of academia actively transferred into industries, France is driven by major players in their own field: BULL for the hardware and cloud solutions, Dassault Systèmes and ESI for the software, TOTAL, EDF, Airbus, Renault, PSA and many others for the applications. SMEs and other domains (pharmaceutics, banking) are following in their footsteps, thanks partly to many national and European initiatives, so that France will keep increasing its footprint on HPC.

REFERENCES

CCRT report "10 ans au service de la recherche et de l'industrie" – 2013.

EESI deliverable "D3.3 Working Group report on Industrial and Engineering Applications: Energy, Transports." December 2011. http://www.eesi-project.eu/pages/menu/project/eesi-1/publications/working-group-reports.php.

IDC Study "Creating economic models showing the relationship between investments in HPC and the resulting financial ROI and innovation—and how it can impact a nation's competitiveness and innovation." October 2013, IDC #243296, Volume: 1.

IDC Studies "A Strategic Agenda for European Leadership in Supercomputing: HPC 2020" and "Financing a Software Infrastructure for Highly Parallelized Codes." September 2010, IDC #SR03S.

Top500 list: http://www.top500.org/list/2013/11/.

Xerfi Study "Le marché du calcul intensif et de la simulation numérique: nouveaux usages et débouchés, prévisions 2015, analyse du paysage concurrentiel." November 2013.

CHAPTER 5

Industrial Applications on Supercomputing in Germany

Michael M. Resch and Andreas Wierse

CONTENTS

5.1 HISTORY

Computing has a long history in Germany and is rooted deeply both in research and in industry. Two of the pioneers of supercomputing are German. David Hilbert (1862–1943), a prominent mathematician, gave an overview talk about the key problems of the coming twentieth century during the World's Fair in Paris in 1900. Among the topics he mentioned were the key questions for the numerical solution of partial differential equations that were the basis for many of today's simulations both in research and industry. Hilbert was also an outstanding teacher. In the years after his talk in Paris, a number of excellent scientists joined his institute in Göttingen. Among them were Richard Courant, who was his PhD student, and John von Neumann with whom he worked on quantum mechanics. The second outstanding supercomputing pioneer was Konrad Zuse (1910–1995). In 1939, he built his first computer (Z1), and, in 1941, he built Z3, the world's first programmable computer.

The history and career of both men highlight the strengths and weaknesses of German supercomputing. David Hilbert is a representative of the German mathematical school that very early focused on application-driven problems and has continued to do so until today. However, between 1933 and 1945 many of the most brilliant scientists had to leave Nazi Germany—a brain drain that damaged Germany's science substantially. Konrad Zuse's invention was not recognized as being relevant by Germany's public authorities, then both the Z1 and the Z3 were destroyed during the war. Although a number of German electronics companies, such as Telefunken and Siemens, actively worked in the field, German hardware industry never again achieved a relevant worldwide position.

However, both David Hilbert and Konrad Zuse still stand out in the history of supercomputing in Germany. First and foremost, industrial use is connected closely to scientific progress. German industry and German research organizations traditionally cultivate a close collaboration. This is also seen in the use of supercomputing.

5.1.1 Industrial Take-Up

Early industrial adopters of supercomputing were found mainly in three different fields. The importance of supercomputing became evident in nuclear power plant engineering. To the extent that nuclear power became important for energy supply in Germany in the 1960s, companies started to use supercomputers for the simulation of critical processes. The use of large-scale systems included the development of software. Nuclear power plant simulation has remained an important topic to this day. Even though Germany decided to abandon nuclear power as a source for energy supply after Japan's Fukushima Daiichi incident of March 11, 2011, supercomputing still is an important tool in this field.

Aerospace engineering was the second field in which supercomputing was used early on. Companies such as Dornier and MTU started to use systems for the simulation in aerospace applications very early. In 1989, DASA (Deutsche AeroSpace Aktiengesellschaft) was founded that continued to use supercomputers in a variety of fields such as computational fluid dynamics, mechanics, and electromagnetics. Today, DASA is part of EADS (European Aeronautic Defense and Space Company) and continues to use supercomputing heavily.

Automotive engineering started to use supercomputing in the 1970s. Leading automotive companies such as Daimler-Benz, BMW, and Volkswagen started to investigate the potential of supercomputing for a variety of fields. Although computational fluid dynamics was adopted early both for external and for internal flows, crash simulation was more difficult to integrate as it was less accurate and reliable compared to computational fluid dynamics. When supercomputing capability increased, the teraflop-barrier crash simulation also became much more relevant for the development process.

5.1.2 Supercomputing in Research

Supercomputing was both a research activity and a tool to support research since the early 1960s. In the beginning, research organizations and universities followed individual strategies. Supercomputing was hence mainly available in places where traditional use like computational fluid dynamics was strong. As a result, large-scale systems were found early on at technical universities such as RWTH Aachen or the University of Stuttgart.

This situation changed in 1995. The German Science Council adopted a national strategy for the supply of supercomputing systems to the German scientific community. Part of that strategy was the creation of national supercomputing centers. The first such center was established in 1996 in Stuttgart—the Höchstleistungsrechenzentrum Stuttgart (HLRS). The Jülich Supercomputing Center (JSC) and the Leibniz Rechenzentrum (LRZ) in Munich (today Garching) followed so that by the end of the 1990s German research had access to three large-scale competitive systems.

In addition, topical centers had evolved over time. Specifically important are the German Weather Service (DWD at Offenbach) and the

German Climate Research Center (DKRZ at Hamburg). Although DWD uses leading edge systems to guarantee high-quality weather forecasts, DKRZ provides access to a large-scale system and large-scale data management for the German climate research community.

In 2007, German supercomputing entered a new era. Driven by European competition and the need to compete worldwide, HLRS, JSC, and LRZ decided to join forces and created the Gauss Center for Supercomputing (GCS). GCS was able to establish a leading role in supercomputing worldwide. At the same time, technical services and support for German researchers were unified and optimized. Today, GCS takes a leading role in the supply of systems, services, and training. In 2008, the Gauss Alliance (GA) was formed, bringing together GCS with a number of smaller centers.

5.2 CURRENT SUPERCOMPUTING SERVICE PROVIDER LANDSCAPE

As we have seen, the German supercomputing landscape has evolved continuously over the last decades and is constantly changing. Here, we distinguish between service providers in the public space and service providers for the industrial area.

5.2.1 Supercomputing Service Provider Landscape for Research

Supercomputing services for research are currently undergoing a substantial change. The institution-based approach that was established in the early 1990s is gradually being replaced by a national concept that is based on two main principles.

1. A pyramid of performance which assumes that
    a. At the Tier-1 level, there is a small number (2–4) of national high-performance computing centers
    b. At the Tier-2 level, there is a larger number (6–8) of regional high-performance computing centers
    c. At the Tier-3 level, there is a very large number of low-end systems—typically clusters—which are available to support basic requirements
2. Bridge the gap between various research organizations such as universities and national research centers and create a unified supercomputing landscape (Figure 5.1).

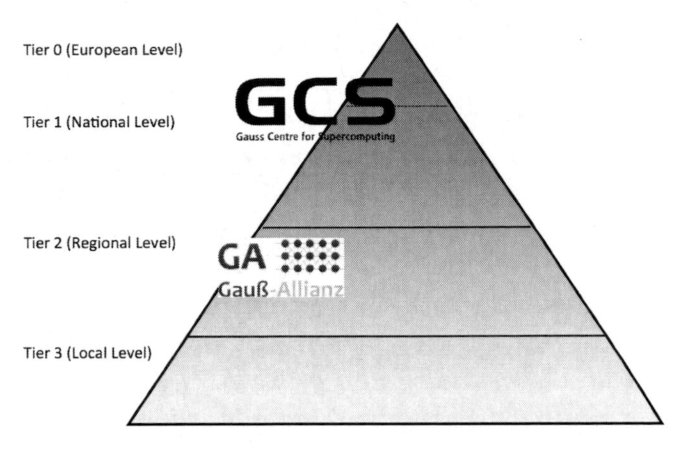

FIGURE 5.1 German pyramid of performance.

At the Tier-1 level, we find the GCS as described earlier. Together, the three centers provide services to all German researchers that require access to high-end systems. Access is controlled by a scientific steering committee, which awards CPU time based on proposals submitted through a portal. Services of GCS are provided in the following fields:

- Access to high-end systems
- Support for porting and optimization
- Education and training
- Support for industrial use of supercomputing

GCS has also established working groups to discuss and agree on major questions that affect the users of all the three centers. Such working groups include the following:

- Operational issues
- Security issues
- Industrial use issues
- Administrative issues
- Public relation issues

By establishing a director and a common office, GCS is able to bundle competencies and skills in a single organization. At the same time, each of the three centers is focused on individual strengths. Together, they cover the whole landscape of application fields and most methodological aspects of supercomputing. Furthermore, at the national level, GCS helps to communicate with political decision makers and scientific funding agencies.

At the same time, GCS acts as a European provider (Tier-0) in the frame of the European PRACE (Partnership for Advanced Computing in Europe) initiative. By providing CPU cycles to European scientists, GCS plays a major role in supporting pan-European activities and projects. At the same time, their role as European provider has helped GCS centers to extend their reach into other European countries and gradually develop new services and methods on an international scale.

At the Tier-2 level, we find a collaboration of medium-sized centers. The GA was established 1 year after the GCS and brings together centers that reach out beyond their home organizations to support a regional user community. GA has started to coordinate these centers and is to establish a central office in the near future.

GA brings together both regional and topical centers. This means that the needs of the centers and their users are more diverse than is the case for GCS. Hence GA is working out methods to intensify collaboration and to provide all users in Germany with the same level of support. Both GA and GCS can base their collaboration on projects that developed basic technologies for distributed resources. One such project was the UNICORE initiative that developed a portal solution for access to supercomputing systems. Another key project was D-grid, the German grid initiative that brought together a large group of centers and developers and provided its partners with a variety of solutions for grid computing. Based on such solutions the target of GA is—by closely working together with GCS—to be able to provide Tier-2 capacity to all researchers in Germany.

5.2.2 Supercomputing Service Provider Landscape for Industry

In the industrial field supercomputing services are either internal or external services.

Internal services typically exist in large-scale companies such as the automotive, aerospace, and pharmaceutical industry. Most of these companies have gradually developed their IT and have developed smaller or larger supercomputing departments to give support for simulations. For most such companies in Germany, supercomputing is done in a central department: These departments typically make an annual capacity planning collecting the requirements of their users and calculating the average computational need. Given the centralized nature, problems such as security and load balancing can be handled rather easily. There are, however, large-scale companies in Germany that rely on decentralized in-house solutions. In such cases, individual departments—operating as profit centers—require supercomputing as a part of their business and hence operate their own machines. These in-house solutions should not be ignored when discussing the landscape of industrial supercomputing as they are potential players in the field and might be outsourced with the next wave of cost saving in large-scale companies.

External services have been growing gradually over the last years. The key issue in this field was—and still is—confidentiality of data and network connectivity. Basically in this field we see two different types of solutions.

First, there is the compromise solution: systems are still inside the company but staff is coming from an external service provider. All technical decision making is done by the external service provider. All requirements come from the user company. As systems are inside the company, neither security nor networking are real issues. Competence for supercomputing can be kept in-house if in such a situation internal and external experts work together and keep a constant flux of knowledge and information. The German startup science + computing—a spin-off of the University of Tübingen—is very successfully offering such services mainly for automotive industries. Having established itself as a provider in a number of German automotive companies, it was recently acquired by the French company, BULL.

Then there is a growing market for outsourcing supercomputing. Such outsourcing requires a substantial financial input before services can be provided. Hence it is not surprising that the number of providers is small and that services are often based on individual contracts rather than on public offerings. In the TOP500 we find only two companies from Germany listed as IT-providers. The most interesting one is T-Systems, which offers supercomputing services to both industry and research. In 1999, T-Systems together with the German Center for Aerospace (DLR) founded the subsidiary T-Systems Solutions for Research (SfR) to provide IT services—including supercomputing services—to DLR.

5.3 GERMAN SUPERCOMPUTING ECOSYSTEM

Funding for supercomputing in Germany goes through a variety of channels and has undergone major changes over the last years. For decades, supercomputing systems were funded cooperatively through federal and state governments. With the change of the German constitution in 2006, this concept was limited to special cases. Based on the new article 91 of the constitution, co-funding is only possible for projects that are of general supraregional (potentially national) relevance. This is a problem for supercomputing as supercomputers generally are not used for a single project. As a result, Germany has seen an intensive discussion about the future funding of supercomputing—bringing together issues like the general role of federalism in Germany with the question of where in Germany large-scale scientific instruments should be placed. The German science council (Wissenschaftsrat) initiated a working group in 2013 to propose a new concept for the future funding of supercomputing in Germany. The working group is expected to come up with a concept by mid-2014.

The following paragraphs describe four basic funding concepts:

Federal research centers: These are usually large-scale institutions where simulation plays a key role and hence supercomputing is a major requirement. The funding for such federal centers is primarily based on federal funding with a small fraction of funding coming from the hosting state. These internal supercomputing centers typically get the funding for their systems from the budget of the federal center or they apply for additional funding. They will not be further considered in this description.

Tier-1 GCS funding: GCS is currently funded through a project. The funding of 400 million euros over 6 years is shared between the federal government and the hosting states of the three centers involved. The federal government covers 50% of the costs while the three hosting states (Baden-Wuerttemberg, Bavaria, North-Rhine-Westphalia) cover the other 50%. Future funding might be through a project again but a political decision has not yet been made and is expected in the year 2014.

Tier-2 GA funding: Funding for the regional centers in the GA comes through the general funding model for co-funded activities. Given the problems with the constitutional regulations as described above, a special line of funding was established, which is supposed to provide €20 million per year for the coming 5 years. Fifty percent of this funding again comes from the federal government.

Tier-3 local systems funding: Tier-3 funding is usually state-based.

5.4 GERMAN SUPERCOMPUTING INDUSTRY

The German supercomputing hardware market faces a significant lack of manufacturers of supercomputing systems. The only company that manufactures supercomputing systems on a large scale that are available beyond the borders of the German speaking countries is Fujitsu Siemens with their production site in Augsburg.

However, there is a whole set of companies that consider themselves manufacturers, but could better be classified as assembler or integrator. Among them are Delta Computer Products GmbH in Hamburg, ICT (information communication technology) GmbH in Aachen, MEGWARE Computer Vertrieb und Service GmbH in Chemnitz, and last but not the least, Transtec AG in Tuebingen. MEGWARE is the only company with a supercomputing system to show up in the TOP500 list in November 2013.

On the software side there is much more activity. Companies are active mainly in two fields. On the one hand there are companies providing tools and middleware for supercomputing. On the other hand Germany has a long tradition of developing application software.

Tools and middleware development benefitted from the SUPRENUM project that started in 1985 and ran for 5 years. Companies such as Pallas and GENIAS took a leading role in the 1990s but ended up being acquired by Intel and Sun, respectively. Today, software is developed in the Fraunhofer society with new solutions being made available to the market regularly. One good example is SCAPOS—founded in 2009—which provides solutions both in tools and applications.

A German world leader in its field is Vampir. Originally developed by a group led by Professor Wolfgang Nagel at Jülich, it is now supported and marketed by a subsidiary of the Technical University of Dresden. Vampir is widely used in the United States, Asia, and Europe.

On the other hand there is a variety of mostly small companies that develop and support supercomputing application software. Most of them are focusing on aerospace and automotive applications. As there are so many, we can name but a few, such as FE-Design, LASSO, Intes, SFE, FluiDyna, and T-Systems. These companies also provide pre- and postprocessing solutions and services in supercomputing of all kind.

5.5 CASE STUDIES AND BUSINESS CASES

5.5.1 Public–Private Partnership

In German research systems, collaboration between industry and public research organizations has always been strong. In the field of supercomputing, such a public–private collaboration was especially close at the University of Stuttgart. Already in 1986, when a Cray-2 system was installed at the University, the computing center started to work closely with Porsche. The field of interest was mainly computational fluid dynamics. Later on also crash simulations became important.

In 1995 the State of Baden-Württemberg, the University of Stuttgart, Porsche, and Daimler took the next step in the public–private cooperation. Triggered by discussions about the optimum use of resources the partners agreed to establish a public–private partnership called HWW (Hoechstleistungsrechner fuer Wissenschaft und Wirtschaft GmbH/ Supercomputing for Academia and Industry Ltd.). The purpose was to bundle resources and demands of the academic and industrial partners and provide the necessary legal and organizational framework to leverage expertise and funding. While initially the company was focused on the operation of supercomputing systems, it gradually shifted its focus to organizational issues. The group of computing systems that was brought together in 1995 gradually turned into a production cloud. Once operational issues were resolved, flexible use, service level agreements, and ease of access became important. At the time of writing, HWW distributes about 120 million core hours per year, mainly coming from HLRS.

Recently, HWW was remodeled to allow for service provisioning based on cloud concepts in the field of simulation and development cycles. Services are offered to the general public with a focus on industrial customers. The rules behind the HWW infrastructure and its service offering are very strict. From the beginning there was a strong security concept which was updated in 2013 to bring security needs and current technology together. Every system made available via HWW has to follow these security guidelines that have been developed by experts of the involved shareholders (in this case University of Stuttgart, Karlsruhe Institute of Technology [KIT], Porsche, and T-Systems); and currently a project is running to integrate all HWW systems into a common user platform, thus making it easier for the users to use the different computers.

Collaboration between public research and industry is driven by a number of factors. Changes in industry are perhaps more rapid than in the public sector. However, the cloud discussion has unleashed a wave of change also in the field of public research computing services. The key questions to be discussed for the future are the following:

- What is the potential need for industry that could be covered by public research?
- If public research and industry can benefit from mutual use of services, what should be the collaboration model?
- And finally, if such collaboration is to work, what are the main problems and roadblocks to overcome?

One argument goes that industry has a substantial need for computing power, which moreover varies substantially over time. The solution to this for industry, it is claimed, would be to keep basic computing resources in-house and access academic resources mainly in times of peak demand. Academia, on the other hand, could provide such peak load without compromising academic research requirements. This argument assumes two things:

1. Industry is interested in outsourcing computing activities to begin with.
2. Academia can more easily postpone its own computing needs to provide access to its systems.

The key question to answer is this: Are companies interested in using academic resources in case of insufficient resources in their enterprise?

5.5.2 Porsche: A Success Story

One of the strongest users of the HWW systems is Porsche, especially the development branch of Porsche located in Weissach. The engineers have been using supercomputing systems for a long time, the cooperation between Porsche and the predecessor of HLRS at the Stuttgart University started in the 1980s. Today, Porsche has its own data link to the premises of the HLRS in Stuttgart and has outsourced its high-end computing needs almost completely to HWW. Every car model that has been developed in the past years—including the famous Panamera—had its crash simulations running on the HWW systems as well as its aerodynamics simulations and its engine simulations.

Porsche has, for example, integrated the HWW systems into the everyday workflow of its simulation engineers, so that it is almost fully transparent to them. HLRS in turn has taken into account the needs of Porsche when it comes to systems architecture as well as security. However, HLRS has not forgotten that it is principally a research computing center and the satisfaction of its academic users must not suffer from its industrial activities. The whole computing ecosystem has been finely balanced in such a way that industrial users find an environment that provides a highly productive work flow, whereas the academic users benefit from the improved service as well as from the revenue stream (all money that comes in via HWW will be reinvested into new hardware, which naturally will also be used by the scientists).

5.5.3 General Business Case

Looking at this example and at the success of simulation in engineering design, it is obvious that there is a true need for simulation in a large group of industries. Even for small- and medium-sized companies, simulation has become a standard tool. More and more large companies are outsourcing development tasks and request that simulation be used to verify design parameters. So we can expect companies of all sizes to have an interest in using cloud resources. For larger companies, the driving factor might be overflow capacity while for small- and medium-sized companies the HWW cloud could be an enabling technology, which allows them to compete with large companies without having to invest heavily in computing infrastructure. In addition, both groups can benefit from the access to a high-end system that allows them to compute models that are so large they could never afford to own a system big enough to run it.

For small- and medium-sized enterprises (SMEs) slow network connectivity and lack of technical skills inhibit the use of external computing resources. Although network connectivity will improve in the future, lack of specialized technical skills might remain an inhibiting factor.

5.5.4 Small- and Medium-Sized Enterprises

Especially in the state of Baden-Württemberg, there is a strong focus on SMEs. Although large companies such as Bosch, Daimler, or Porsche are well known, there is a sizeable number of smaller companies around them, which not only support them as suppliers, but are also active in many different business fields. We illustrate this by presenting two examples that show how different they can be.

5.5.4.1 Large-Scale Power Plant Simulation

The first company we will look at has its roots in the University of Stuttgart. Scientists at the Institute of Combustion and Power Plant Technology (IFK) developed a simulation code that allows computing the complex processes that take place in the combustion chambers of large furnaces or boilers. Their focus is on problem analysis, design, and process optimization. The development of the simulation code RECOM-AIOLOS started at the institute and is now continued by RECOM Services GmbH.

A team of 10 people around the founder and managing director Dr. Benedetto Risio is using RECOM-AIOLOS to perform simulations for companies that own or build large industrial furnaces. For the operators of combustion chambers and manufacturers of plants, it is extremely important to understand the risks in the process early on and thus achieve better control of the costs. Their customers often have their own teams for simulation, but the mapping of the overall process is too expensive for them and requires specially trained staff.

To assure that the results of the simulation in the virtual reality match the real world, the starting point is to compare a certain number of operational states by looking at the measured and simulated values. It only makes sense to, for example, vary the components of the fuel or modify the burner and come to reliable results when a significant coincidence occurs.

Because of the limited amount of internal computing power, it is extremely valuable for RECOM Services to be able to access the large computers at the HLRS. Especially when performing their research to assure their business of tomorrow, the access to the supercomputing center is crucial. Currently they are working on the development of genetic algorithms for the automated optimization of design and operation of industrial boilers.

However, the use of the supercomputing center is not only limited to raw compute power. It is equally important for RECOM Services to explain the results to their customers in such a way that they really trust these results enough to base decisions for expensive measures on them. Given the complexity of the processes, visualization is by far the most efficient way to communicate the findings. The five-sided CAVE at the HLRS is the ideal environment for both analysis of the results by the simulation experts and communicating these results to the customer and visualization, being directly integrated into the high-speed backbone of the computing center, to directly access the results and bring them to life in the virtual environment. And it is here in the CAVE where decisions are made.

5.5.4.2 Simulation of Production in Clean Rooms

The second company comes from a similar field, but takes a different approach. Optima Pharma is located in Schwäbisch Hall and employs about 550 people. Their business is to develop, design, and build machines and manufacturing plants for aseptic and sterile packaging and filling technology in human and veterinary medicine. The emphasis is currently on air flow required in cleanrooms, to make it valuable as a quality factor, and to optimize it.

Not too long ago, cleanrooms had to be built physically several times, to measure and optimize the air flow. Those days are over at Optima Pharma. A new engineer in the process technology development group brought with him the idea to use simulation technology. Supported by workshops and by using software test licenses, Optima Pharma first developed a simple smoke tube simulation. This simulation showed that a smoke tube can disturb the flow in the cleanroom due to vortices and the displacement of air around the smoke outlet.

The next step aimed to simulate a cleanroom including the filling machine—and to visualize the results appropriately. The crucial impulse and the necessary support came from HLRS in Stuttgart. HLRS provided the opportunity to use suitable open-source software and access the required computing capacity. Today, Optima Pharma is able to perform the desired simulation and to extensively visualize the results in the 3D Virtual Reality environment.

The most important benefit for Optima Pharma lies in significant time and cost savings. The knowledge that can be acquired through simulation well before the physical construction of the plant significantly minimizes the risk of incorrect planning. And to see their future plant in the 3D environment is a great experience for the customers—a very descriptive one as well.

The next steps for Optima Pharma will lie in the field of thermodynamics and multiphase flows. They see a large potential in the use of simulation and visualization technology.

5.5.5 Stimulation Plans for Adoption of Supercomputing by the German Industry and SMEs

Among the German supercomputing centers, the HLRS in Stuttgart has by far the longest track record with respect to cooperation with industry and especially SMEs. The HWW GmbH mentioned above has only recently opened its systems to not only serve its shareholders but also third parties. And the University of Stuttgart together with the Karlsruhe Institute of

Technology (KIT) has founded Simulation Computing and Storage Ltd. (SICOS BW GmbH) whose main purpose is to improve the availability of supercomputing, especially for SMEs.

Financially supported by the Ministry of Science, Research and Art of Baden-Württemberg, SICOS has been founded to achieve the following:

- Inform enterprises with emphasis on SMEs about the possibilities of simulation on supercomputing systems and their availability for commercial users
- Support them in their first stages such as defining test cases, performing pilot projects, and gaining trust in these new methods (new for them)
- Find the right partners that are able to help them solve their application problem (engineering services, research institutes, independent software vendors, etc.)
- Optimize the internal processes inside the supercomputing centers with respect to the needs of commercial users (easing administrative processes, communication, support technical process improvements, etc.)
- Be a link between the needs of the industrial users and the capabilities of the supercomputing centers (harmonize security concepts, co-ordinate user management, etc.)

Application knowledge is a very critical aspect when companies are looking for simulation and supercomputing. It has therefore been taken into account that this knowledge is usually fairly restricted in the supercomputing centers compared to the wide area of potential application fields in the industry. Baden-Württemberg has therefore developed the concept of solution centers. Today, two HPC solution centers exist:

1. The Automotive Simulation Centre Stuttgart (ASCS)—founded in 2008—that focuses on the cooperation of OEMs, suppliers, researchers, and ISVs in the precompetitive areas.
2. The Energy Solution Centre (EnSoC)—founded in 2009—that has members from energy providers, IT companies, and research institutes dealing with the challenges in energy economics.

Solution centers form a platform that can be used by their members to exchange their views, define common needs, organize groups for proposals for research grants, or team up in projects satisfying their requirements. Each solution center is closely linked to a supercomputing center to bring all important components together. Currently three additional supercomputing solution centers are under preparation: one for chemistry and materials, one for life science, and one for media and art.

All these activities are concentrating on quality, not quantity. A common goal includes the creation of success stories—examples of how SMEs can benefit from the power of supercomputing and simulation. They will in turn enlighten other companies and make them take a closer look at what supercomputing actually is and what it can do for them. The balance between supercomputing providers, end users, ISVs, and service providers is very delicate; true information exchange is only possible in a trustworthy environment.

5.6 CONCLUSION

Supercomputing is deeply ingrained in the processes of research and industry in Germany. With the traditional close cooperation of research and industry systems, methods, applications, and processes are exchanged between the two fields on a regular basis.

Despite having widely ignored the relevance of hardware in the 1980s, today Germany holds a strong position in the use of supercomputers and in the development of algorithms and software. As German companies were early adopters of the technology, use of supercomputers is strong in the economic world. We have shown some examples of larger companies like Porsche, medium-sized ones like Optima Pharma, and small ones like RECOM. Based on supercomputing technology, German companies are setting out to defend and enlarge their worldwide market share in high-technology markets such as production machinery, process engineering, pharmaceutics, automotive, and aerospace. In many other fields, in which German SMEs are hidden champions in specialized niches, supercomputing will help support economic growth and technical excellence by keeping costs low, speeding up processes, and developing new ideas.

CHAPTER 6

Industrial Applications on Supercomputing in Italy

Sanzio Bassini and Claudio Arlandini

CONTENTS

6.1 THE NATIONAL HPC POLICY AND THE LOCAL ECOSYSTEM

At the end of 2007, the working group ESFRI (European Science Forum for Research Infrastructures) identified high-performance computing (HPC) as a strategic priority for Europe. Also because of the implementation costs related to constant growth, research infrastructures, such as the most advanced computing facilities, often require transnational or continental contributions from many member states. A paradigmatic example of this trend is PRACE, the transnational action supported by the European Commission to design and implement the infrastructure for European supercomputing.

Since the very beginning, Italy's presence in this Consortium was considered strategic because it was allowed to participate in the definition of guidelines for the implementation of more powerful computing infrastructure at the European level, thus allowing the country to affect the policies of access and use of resources that would have become available. Thanks to the authority granted by the Ministry of Education, University and Research at CINECA, Italy was able to carry out, in accordance with the evolution of the national scenario, a specific role in this initiative as Principal Community Partner in line with the role of the major European countries.

In 2012, CINECA installed a computing system of class Tier-0, positioned at the top of the TOP500 List of the largest supercomputers in the world.

6.2 HPC NATIONAL ACADEMIC AND BUSINESS SUPPORT CENTERS: INFRASTRUCTURE AND SERVICES

6.2.1 Introduction

The Italian HPC ecosystem saw in the last 2 years a major reconstruction, dictated by the economic situation of the country. The three University Consortia sharing the institutional mission of supporting research and support for technological innovation in Italy—CINECA, CILEA, and CASPUR—merged in a unique entity, maintaining the name CINECA. The aim was to rationalize government expenses and to strategically strengthen the international competitiveness of Italy in the field of research and education by improving the efficiency and synergy between organizations and institutions.

Other minor infrastructure and research centers with HPC-related activities do exist, mainly born in the past years with funding related to specific projects, but their focus is usually quite domain-oriented, and their capability is limited compared to CINECA.

6.2.2 CINECA

CINECA, the largest computing center in Italy, is a not-for-profit Consortium, made up of 69 Italian Universities, two National Institutions, and the Ministry of Education and Research.

CINECA was born in 1969 when, with the support of the Ministry of Education, four Italian universities united their effort to create a common computing center and install the first supercomputer in Italy. Since then, especially since the 1980s, the Consortium has extended its mission by developing institutional management and administrative systems for universities and transferring their skills to the world of public administration, businesses, and services, with a consequent increase in the number of employees and the number of services offered. Today, CINECA is a point of reference for the national academic system, the technological heart of the system of communication between the University and the Ministry of University and Research. In addition, thanks to the experience gained in over 30 years, CINECA has taken a leading role in the field of Public Administration, and its skills are becoming increasingly popular in industry and services companies.

SCAI (Super Computing Applications and Innovation) is the HPC department of CINECA. The mission of SCAI is to accelerate the scientific discovery by providing HPC resources, data management and storage systems and tools and HPC services and expertise at large, aiming to develop and promote technical and scientific services related to HPC for the Italian and European research community.

CINECA enables world-class scientific research by operating and supporting leading-edge supercomputing technologies and by managing a state-of-the-art and effective environment for the different scientific communities. The SCAI staff, comprising about 150 technicians, offers support and consultancy in HPC tools and techniques and in several scientific domains, such as physics, particle physics, material sciences, and chemistry.

CINECA is currently recognized as

- A Compute Unified Device Architecture (CUDA) Research Center, based on the vision, quality, and impact of its research leveraging Graphics Processing Unit (GPU) technology

- An Intel® Parallel Computing Center, to accelerate the creation of open standard, portable, and scalable parallel applications by combining computational science, hardware, programmer tools, compilers, and libraries, with domain knowledge and expertise
- The main development pole for software tool Quantum Espresso, an integrated suite of Open-Source computer codes for electronic-structure calculations and materials modeling at the nanoscale[1] and SPECFEM3D, a tool to simulate acoustic (fluid), elastic (solid), coupled acoustic/elastic, poroelastic, or seismic wave propagation[2]
- One of the six PRACE PATCs (PRACE Advanced Training Centres), to carry out and coordinate training and education activities that enable the European research community to use the computational infrastructure available through PRACE

6.2.2.1 CINECA Infrastructure

CINECA is currently one of the large-scale facilities in Europe and PRACE Tier-0 hosting site.

Users can find a catalog of software programs for different jobs and disciplines, a powerful and efficient network for local and external connection and a large number of experts with a wide range of skills and knowledge that can help researchers, both from the technological and scientific point of view.

The largest CINECA system is FERMI[3] (Figure 6.1), an IBM Blue Gene/Q supercomputer classified among the most powerful supercomputers in the TOP500 List: rank 7th in June 2012. FERMI is comprised of 10,240 PowerA2 sockets running at 1.6 GHz, with 16 cores each, totaling 163,840 compute cores and a system peak performance of 2.1 petaflops. Access to FERMI is on the basis of national and European calls for proposals. ISCRA (Italian SuperComputing Resources Allocation) and PRACE manage the access to the Tier-0 supercomputer by the way of international peer-review procedures ensuring world-class research.

The recently installed prototype Eurora,[4] born by a joint effort of Eurotech and CINECA, is based on a hot water cooling system, delivering 3,209 megaflops per watt of sustained performance, and is one of the world's greenest supercomputers: 1st in the Green500 List of June 2013. Its energy efficiency improved upon the previous greenest supercomputer in the world by nearly 30%.

FIGURE 6.1 CINECA FERMI BG/Q system.

Each one of the 64 Eurora nodes is equipped with two Intel Xeon Sandy Bridge processors and either two NVIDIA K20 accelerators or two Intel Xeon Phi™ 5120D coprocessors.

A third system, the PLX cluster,[5] is especially devoted to industrial services and specific scientific projects developed in collaboration with CINECA specialists. It is equipped with 274 nodes, equipped with 2 six-core Intel Westmere 2.40 GHz per node (548 processors, 3,288 cores in total) and 2 NVIDIA Teslas per node for a total of 548 GPUs.

PLX will be replaced by a new system in the second half of 2014, aiming to a peak performance of at least 1 petaflop.

6.2.2.2 CINECA Services

CINECA makes a clear distinction between services for academic research with respect to industrial research. For academic research, access to HPC resources is free on the basis of an open-access peer-review procedure.

Industry may obtain HPC cloud services on a commercial basis on the PLX cluster, mainly through Super Computing Solutions (SCS*), a company totally owned by CINECA. SCS provides a wide offering of services, both for companies that have their own HPC resources in terms

* http://www.scsitaly.com/.

of small cluster and companies that do not want to buy or manage their own resources. Data Security, Service Flexibility, Remote Visualization, and hardware performance are the key features of SCS Cloud service. Customers who need fast access to resources for peak production can set up their environment and their application, test production, and get a defined amount of dedicated computing resources when they actually need them. Consulting activities on hardware, system configuration and technology trends are available for those companies that want to buy their own HPC systems. SCS is available for Independent Software vendors (ISVs) SW testing and also Computational Fluid Dynamics (CFD) consulting based on OpenFOAM, an open-source CFD application.

The activity of technology transfer is one of the main missions of CINECA and of SCAI department in particular. A range of services are at the disposal of industries that can benefit from HPC, on the basis of a long tradition of high-level quality of services, such as

- Access to state-of-the-art systems most suitable for technical computing
- Advanced and deep technical expertise
- Codevelopment of customer-specific applications
- Continuous support for the entire process or for the project life-time in case of services not included in the catalogue

The use of methodologies of numerical simulation as support of the design process is sometimes stopped or impeded by the costs of software licenses and by the need of computing resources. SCAI is acting as facilitator for large and small/medium enterprises, and its Innovation and Technology Transfer actions toward Small- and Medium-Enterprises (SMEs) is highly watchful to the constantly evolving scenario.

6.3 INDUSTRIAL (POTENTIAL) USERS: EARLY ADOPTERS AND PROMISING MARKET SEGMENTS

Starting from its institutional role, CINECA is fully engaged in the well-known virtuous interplay loop existing between academia, research centers, and industry. The collaboration with industry and the delivery of services to industry complement and reinforce services offered to the academia and to the research centers in a feedback loop improving both. Working with industry requires, besides technical excellence, a strong commitment and focus on production, on the quality of service offered and its continuous improvement, the definition and fulfillment of the service level agreements, the achievement of the project goals in the defined timeframe and within the budget, the point of view on technology trends related to the specific application or business case. On the other side, working with academia allows to look ahead for technology trends considering a wide range of applications and to experiment new programming models or hardware prototypes on a wider base of applications. It allows the testing of solutions that could be affordable in an industrial production in advance developing competence and experience that can be transferred to industry. European Commission as well as the Council of Competitiveness in the United States consider HPC one of the drivers for innovation. This is clearly understood by large companies but less affordable for SMEs. For this reason, the list of early adopters of CINECA HPC-Cloud services is mainly large companies and/or R&D groups working in highly competitive technological fields of application.

To these users, CINECA offers proofs of concept and services based on software applications used in different industrial sectors such as automotive, finance, pharmaceutics, oil and gas, shipbuilding predominantly, computer-aided engineering (CAE) applications. Following are more meaningful details and experiences performed with early adopters and the market segments involved.

6.3.1 ENI

Eni S.p.A. is Italy's largest industrial company and one of the largest companies in the world working in the oil and gas sector. To sustain exploration and production activities using HPC systems, Eni has established a long-lasting partnership with CINECA. CINECA offers and develops a wide range of HPC services for the company. Moreover, both collaborate to test new technologies, to evaluate proprietary applications performance on new hardware, and to train ENI specialists to most effectively take advantage of the HPC infrastructure. The three main specialized services in synthesis are:

1. The daily management of Eni HPC infrastructure
2. On-site support to users during production and management of daily needs
3. Development and maintenance of HPC applications in production

CINECA won an Eni innovation prize in 2009[6] for the development of the DVA (Depth Velocity Analysis) application tool, the engineering of a company-proprietary technology for the reconstruction of seismic velocities and other geophysical parameters in complex geological bodies.

6.3.2 Dompé pha.r.ma

Dompé pha.r.ma, a spin-off of Dompé S.p.A., was created in October 2005 as part of the reorganization of the Group's activities as an independent company, totally dedicated to the activities and processes of research, development, and production of medicines in the areas of oncology and immunology. Dompé pha.r.ma can boast a broad portfolio of skills and ownership structures to support the process of development of new drugs (Drug Discovery). They study in particular monoclonal antibodies, drugs of low molecular weight and recombinant DNA from early stages until clinical development. In the process of discovery, the construction of libraries of compounds, which are focused toward specific cellular targets, combines the use and treatment of many physical and chemical parameters of the compounds and the protein receptors.

CINECA has supplied its HPC infrastructure and know-how of its developers for parallelization and implementation of the simulation model of the drug–receptor interaction. The main output of the project was the development of Ligen, a suite of software for CADD (computer-aided drug discovery) (Figure 6.2) customized on the requirements of Dompé pha.r.ma.

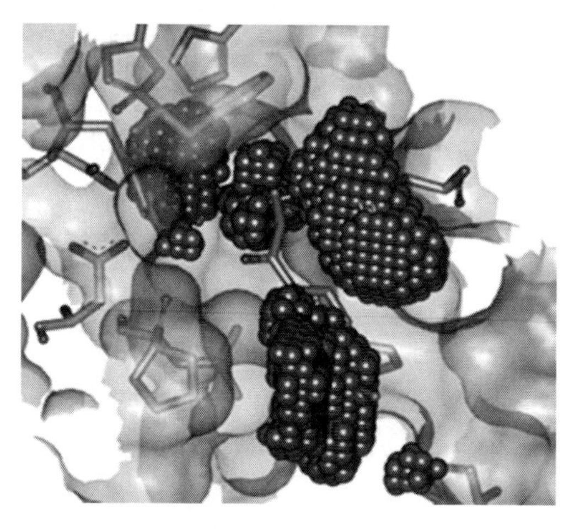

FIGURE 6.2 Active site determination and visualization using Ligen software.

FIGURE 6.3 **(See color insert.)** Luna Rossa AC72 "flies" over the water during a training session.

6.3.3 Luna Rossa Challenge

To win the America's Cup, perhaps the most coveted sailing trophy, one must not only defeat the other teams in regattas but also in design studies, in a technological challenge, in which a major component is computational fluid dynamics (CFD). CFD software open-source, or in any case free of direct costs related to the purchase of license, guarantees cost benefits and flexibility of the software, but in any case requires significant investments in their validation for industrial use cases. The problem is complex in its entirety and it is crucial to have advanced computing resources, over which computational tools can process the large amount of data produced by a similar design study considered as a whole and each single numerical simulation. For this purpose, the design team of Luna Rossa Challenge (Figure 6.3), besides taking advantage of large amount of CINECA HPC resources for design data production, both in terms of computing and remote visualization, has undertaken with CINECA a 12-month feasibility study for the evaluation on the same analyses of the OpenFOAM (Open Source Field Operation and Manipulation) library on HPC platforms.

6.4 PILOTS, LESSONS LEARNED, DOS AND DON'TS, AND SUCCESS STORIES

6.4.1 Introduction

This section presents an overview of recent, or even on-going, initiatives of CINECA's partnership with industries, along with a number of lessons learned in 30 years of history of providing HPC-related services, and success stories.

6.4.2 Pilots

6.4.2.1 The Italian Contribution to SHAPE

The SME HPC Access Programme in Europe (SHAPE) was outlined in early 2013 in the framework of the PRACE-3IP Project,[7] aiming to equip European SMEs with the awareness and expertise necessary to take advantage of the innovation possibilities opened by HPC, increasing their competitiveness.

The main area of the Program's operations is work on a one-to-one basis with SMEs willing to adopt a new, HPC-supported solution. The Program takes such customers as far as trying out the solution on PRACE's infrastructure. The existing solutions such as the PRACE Open R&D Access Programme, the PRACE Advance Training Centers, and the PRACE Project's Open Source code enabling activity are used to provide some of the Program's services.

To evaluate various solutions in the process of the practical project implementation, PRACE ran a Pilot Project between June 2013 and May 2014.

The pilot started with an open Call for Applications to all European SMEs having a business project that can be implemented using HPC. The Call had a brilliant response, involving 14 SMEs, from 7 different Countries (Bulgaria, France, Germany, Ireland, Italy, Spain, and the United Kingdom) and spanning across various industrial domains. Although the Call was not widely publicized, due to the necessity to maintain a limited scope, the response was particularly relevant especially in Italy, because of the 14 SMEs, six came from this country.

The review process was completed in October 2013, when the PRACE SHAPE experts selected ten Projects. Because four of the selected projects were Italian, this also confirmed the quality of the proposals prepared with the support of CINECA.

The selected Italian SMEs presented a wide range of applications and requests. They were:

- ***THESAN S.p.A.*** Thesan is an Italian SME involved in energy sustainability. The project aims to optimize the design of a volumetric machine (hydraulic turbine), improving the overall performance by designing and realizing rotating chambers where the fluid flows, drastically cutting down the time and resources needed to construct a physical prototype and actually building the preferred structure only.

- ***AMET S.r.l.*** AMET is an Italian high-tech engineering company, active in the design and development of mechanic and mechatronic products and processes. The project aims to analyze state-of-the-art simulation techniques to replace the standard deterministic approach used to evaluate the passive safety performances of a vehicle when using composite materials, with a robust lean statistical approach.
- ***Lapcos Scrl.*** Lapcos is an engineering firm that adopts virtual prototyping technologies. The proposed project is aimed at building, upon the OpenFOAM library for CFD, a custom, vertical product for automatically design centrifugal pumps, exploiting HPC resources.
- ***MONOTRICAT S.r.l.*** Monotricat is an Italian company that designs an innovative motorboat hull, characterized by its high hydrodynamic efficiency.

The THESAN case is particularly interesting. THESAN is a relatively young company offering the market innovative solutions in the fields of renewable energy and energy-saving systems. The main goal of the project is to obtain an increased efficiency of an innovative hydraulic turbine. Experimental data show that the main cause of the efficiency losses is the dissipative motion of the fluid during its flow through the machine. A significant increase of the efficiency, up to at least 0.95 in the optimal working point from the starting value of about 0.65, is required to make the machine economically viable.

THESAN started adopting a classical product development strategy, building an actual prototype. CINECA is proposing a codesign strategy, where CFD expertise is introduced to the company technicians to perform 3D simulations to analyze the velocity and pressure field of a fluid flowing through the volumetric machine. The modeling should allow then, through an optimization strategy, to study key parameters of the design of the moving chambers in which fluid flows and to identify the optimum conditions. The objective is mainly to achieve a significant reduction of the time required by the design process, evaluating in a reasonable amount of time spent in the simulation activities the performance of a given design of the machine itself. A side effect, but of primary importance, will be the dramatic reduction of the costs involved in the realization of prototypes. THESAN estimated that with an HPC-supported approach it will be possible to cut down costs and time, reducing the number to one single prototype to be realized within a year after the simulation startup, instead of two to three preliminary versions, saving more than 12 months and about €50,000.

6.4.2.2 The Italian Contribution to Fortissimo

Fortissimo (Factories of the Future Resources, Technology, Infrastructure and Services for Simulation and Modelling) is a Factories of the Future EU project started on July 1, 2013, ending on June 30, 2016. It aims to enable European SMEs to be more competitive globally through the use of simulation services running on HPC cloud infrastructure.

A "one-stop-shop" will greatly simplify access to advanced simulation, particularly to SMEs, making hardware, expertise, applications, visualization, and tools easily available and affordable on a pay-per-use basis. In doing this, Fortissimo will create and demonstrate a viable and sustainable commercial ecosystem.

Fortissimo is driven by end-user requirements: approximately fifty business-relevant application experiments will serve to develop, test, and demonstrate both the infrastructure and the "one-stop pay-per-use shop." Twenty experiments have been defined at the start of the project, in heterogeneous fields. Two open calls for participation in 30 experiments are then expected: the first was opened in November 2013 and the second in May 2014.

Of the initial experiments, three were under the responsibility of CINECA, involving five other partners

1. *Exp. 4.05: Cloud-based multiphysics simulation.* Prysmian Group is world leader in the energy and telecom cables and systems industry. Prysmian is experiencing the need to scale-up the current simulations of energy cables and systems to a larger refinement and size, exceeding the computing power available to them. In addition, they aim to enlarge the limits of what they are simulating, requiring new competences and tools. The experiment will provide technicians with a workflow integrating CAE software with pre/postprocessing visualization and open-source tools in a cloud environment, to create innovative products while reducing time-to-market.

2. *Exp. 4.17: Cloud-based computational fluid dynamics simulation.* Koenigsegg is a Swedish SME whose core business is the development and production of high-performance, high-quality, limited-edition

motor vehicles. Konigsegg and CINECA, together with experts from the National Technical University of Athens, collaborated using iconCFD software. Koenigsegg is committed to produce not only high-performance but safe cars, using all the technology available. If massive incompressible aerodynamic applications are difficult to afford for a company like Koenigsegg, compressible applications, like those necessary to study Drag and Lift at very high cruise speeds, are at least an order of magnitude greater cost in terms of computational resources and data processing.

3. *Exp. 4.18: Cloud-based multiphysics simulation.* EnginSoft is an SME consulting company operating in the field of computer-aided engineering, virtual prototyping, and advanced simulation. One of their challenges is the investigation of the behavior of acoustic turbomachinery. Increasingly, high-fidelity solutions are required by potential customers, requiring finer meshes with transient runs, needing a scale-up of computer resources not available to the company or the customers themselves. To overcome these limitations, a customized workflow based on the software ANSYS CFD will be made available on HPC systems via a cloud-based infrastructure, with collaborative remote visualization capabilities. This will allow Enginsoft to be attractive in terms of business and technological perspective for a new segment of customers.

The first open call, with deadline January 2, 2014, was very successful, surpassing project management expectations, especially for Italy. Of 65 applications, 16 involved CINECA as a provider of HPC competence and infrastructure, and 15 applications involved Italian SMEs (20 in total). The fields of application are extremely varied, ranging from environment to rapid prototyping, from CFD to pharmaceutical.

6.4.3 Lessons Learned

Although every company tends to require a customized approach, and this is especially true for SMEs, a few common approaches do emerge.

The first lesson is that for today's industry, a "classical" offer of HPC cycles as the primary selling point makes no sense any more. Reducing hardware costs is a clear motivation for that. Nevertheless, companies soon discover that buying a cluster with the latest hardware technology does not in itself solve their simulation problems, and it even creates new ones.

A better answer lies in offering companies an integrated bundle of services, including computing, software, visualization, and expertise.

The major challenge from the computing center point of view in creating such an integrated bundle lies in software. When commercial software is required by the customer, or by the application itself, usually the license costs outweigh all other components by far. There are two possible answers to mitigate this issue:

1. Opening a dialogue with a commercial ISV to offer new licensing models, more pay-per-use or "cloud-friendly" orientation. This is the approach the Fortissimo Project is pursuing, where a Workpackage led by CINECA is dedicated to that.
2. Fostering the transition from commercial software to an open-source suitable replacement. This is true for fields such as CFD where open-source packages, like OpenFOAM, are reaching the necessary maturity to be applied in industrial environments.

Helping the transition to open-source requires great effort to overcome industries' resistances, and often prejudices, against open-source packages. The hurdles commonly perceived by industry concern the stability of the solvers or the lack of features considered necessary such as effective Graphical User Interfaces (GUIs) but more preeminently the fear of a lack of support or more generally of missing some hidden costs. The computing center impact lies therefore in building first of all a sense of trust with the industrial partner, starting from limited-scope "demo" projects aimed to demonstrate that numerical accuracy and computing times may be comparable with respect to the commercial software they used to adopt, and that CINECA may be as reliable as an ISV in providing the support they need. This requires time and effort, but allows to build long-standing relationships with the customer. At the same time, the computing center must be clear with the customer showing that hidden costs do exist, that often moving to open source does not mean a net saving in the short-medium run, but moving budget from a license fee to personnel and training.

Particularly important is increasing the integration of computing and remote visualization capabilities. CINECA has in production a web-based interface to job submissions and management, with a plugin to implement remote visualization capabilities. The web-based interface is particularly aimed to industrial users, allowing reduction of the learning time and effort necessary to be productive in an unfamiliar computing system and often operative system. The interfaces are easily customizable allowing to replicate the workflow the end-user is accustomed to on their workstation, and the remote visualization plugin allows the launch of the familiar application GUIs as separate windows.

The use of remote visualization capabilities gives value to the user not only because it recreates on a supercomputer a familiar environment, but even more because it allows to perform pre- and postprocessing tasks remotely, strongly reducing the need of transferring data between the data center and the industry site. The network bandwidth available to the Italian SME is on average not so high; therefore, having to transfer large quantities of input and output data creates delays that may easily overcome the benefits of using a remote computing center in terms of reducing the simulation time. The CINECA solution, based on specialized hardware, having not only high-performance graphics cards but also large shared memory RAM, up to 1 terabyte, allows for the opening of meshes of hundreds of millions of elements, and video stream compressing software, which has been tested successfully by users located in the United States and New Zealand. A "plus" benefit of the system is that it allows a collaborative interactive access to an application interface, allowing the possibility of the team working with not only members geographically remote, but also CINECA technicians to set up customized training and support events with customers.

6.4.4 Dos and Don'ts

Probably the most important issue is "don't use the word supercomputing." This is particularly important when dealing with SMEs. It is clear that a "language gap" exist between companies and the computing center, a gap that needs to be overcome if one wants to create a trust relationship. The term "supercomputing" is a clear example. An SME has usually a very specific business problem to be solved when they approach a computing center. Already approaching a computing center is often a step that, surpassing their familiar process organization, requires some effort and internal discussion to start with. Their problem is very specific and mundane. Speaking about anything "super," be it a supercomputer or supercomputing techniques, appears to them only a marketing buzzword.

The focus needs to be user-centric, talking about the expected value and how to reach it, rather than on the infrastructure. This is sometimes difficult for the computing centers, keen to measure themselves only through the semi-annual TOP500 List.

Another relevant issue is "to create an ecosystem." The introduction of innovative processes or products in a company is by definition a source of risk or it is perceived as such, and introducing HPC-enabled techniques is not an exception. The risk is related to the necessary investments that a company has to make to allow innovation, being related either to acquiring infrastructure or the necessary competences, either through new personnel or ad-hoc training. The computing center has to transform itself in a "catalysis agent" for innovation, helping the SME to manage and reduce risk. This may be accomplished creating an ecosystem for innovation, bringing together all available stakeholders, especially Public Administrations and Universities.

Creating initiatives that allow SMEs to test innovation without the (or with a much lower) risk connected to the necessary investments with pilot cases, which help them to measure the expected Returns on Investment (ROIs), and (eventually) provide support to obtain EU or national funding for production implementation, is an extremely powerful tool, as demonstrated by the success of the SHAPE and Fortissimo open calls. This is a win-win situation for every stakeholder, as the final outcome translates in public money creating new job opportunities for skilled technicians coming from universities.

6.4.5 Success Stories

The experience gained through pilots and other regional initiatives made possible to create a portfolio of commercial services based on HPC to be offered to manufacturing companies. The HPC service was focused on CAE applications and hosted on the general purpose Linux cluster PLX, described earlier. The cluster is equipped with the most common and up-to-date CAE software. The service typically does not cover license costs, because ISVs require a direct contact with customers. HPC needs in manufacturing companies are usually varied and often involve two company functions such as R&D and IT departments, which have different and sometimes conflicting (e.g., performance versus security) requirements. However, CINECA sets up a two-sided offer able to satisfy most of its industrial customers: we present a range of success cases that involve the flexibility of the proposed HPC service.

One of our customers is an electrical appliances manufacturer, based in 60 countries and selling its products in over 150 markets. It is a world leader, with sales approaching €11.60 billion (2011) and employing about 58,000 people. They use internal HPC resources for R&D, mainly CFD applications, but due to increased demand for large optimization studies and a reduction of license costs, their computing demand exceeds current installed capacity. These projects are typically short termed—2 or 3 months—and involve a set of resources comparable with that of the customer. Outsourcing to CINECA gave them the ability to meet the deadline for R&D at a cost comparable to the internal one, with an environment that mimics, when it does not outperform, the performances.

A different example is represented by an Italian customer, which designs, builds, and commercializes hoods, fans, and electrical motors. It is a world leader with several production sites worldwide and consolidated turnover of approximately €385 million with more than 3,000 people. The R&D division that designs and engineers the hood components is based in Italy and they use exclusively external resources for HPC, provided by CINECA. In this case, the key factor is the specific expertise in this domain that made CINECA the best partner for completely handling the HPC service. CINECA provides a complete service, including remote visualization and interactive job management via web interface.

Another Italian customer, a market leader in the production of high-end yachts, provides the final example. For these kind of product, CFD studies are a vital stage within the life span of a project. In fact, CFD analyses are carried out to determine the hydrodynamic resistance according to the specific hull design, to foresee the boat trim, specific analysis on particular appendices, maneuverability tests, and external aerodynamics, which traditionally involved in-field tests or naval basin tests that are much more expensive and require longer time to be performed. Hence, CFD on HPC infrastructure can shorten time-to-market and construction all of costs, which yield a more profitable product for our customer. They typically use their internal, low-sized HPC cluster to set up a particular model, and they go for large parallel jobs on CINECA when their analysts completely model the case.

In conclusion, these three scenarios show that an elastic and reliable HPC infrastructure in conjunction with trained and committed user support, delivered with a flexible commercial offering might play a concrete role in this very specific niche of the more generalist IT market.

6.5 CONCLUSIONS

In summary, we consider it essential to stress once again the most important lesson we learned fostering innovation to SMEs. SMEs are not able, with very few notable exceptions, to finance their own innovation. At this stage, having development programs that fund projects directly with SMEs is crucial for promoting innovation through advanced simulation and computing. National and European programs, like Fortissimo, that fund projects directly with SMEs are a great starting point to work as an operational tool. Nevertheless, speaking with industry requires that the discussion be conducted by experts of the application domain and the industrial sector, rather than by HPC experts. In that sense, the action of most of the running projects, Fortissimo *in primis*, needs to be strengthened in this regard.

REFERENCES

1. Quantum Espresso, http://www.quantum-espresso.org.
2. SPECFEM3D, http://www.geodynamics.org/cig/software/specfem3d.
3. FERMI description, http://www.cineca.it/en/content/fermi-bgq.
4. Eurora description, http://www.cineca.it/en/content/eurora.
5. PLX description, http://www.cineca.it/en/content/ibm-plx.
6. ENI Innovation Prize 2009, http://www.eni.com/en_IT/innovation-technology/technological-answers/explore-challenging-resources/explore-challenging-resources.shtml.
7. PRACE Integrated HPC Access Programme for SMEs. *PRACE-3IP Project Deliverable D5.2, 2013.*

CHAPTER 7

Industrial Applications of High-Performance Computing in Japan

Makoto Tsubokura

CONTENTS

7.1 HISTORY OF HPC IN JAPAN AND NATIONAL SUPERCOMPUTING PROJECTS

7.1.1 Hardware Development and Governmental Support

The development of Japanese supercomputers in its earlier days was mainly supported by the Ministry of Economy, Trade and Industry (METI). The IBM System/360, introduced in 1964, had heralded the third generation of computers (computers controlled by vacuum tube were of the first generation and those by transistor of the second generation) controlled by integrated circuits (IC) or large-scale integration (LSI) chips, and it had quickly swept the Japanese market for mainframes. The widespread popularity of this machine impressed on the Japanese government the need to support its computer industry, and in 1966, METI launched a national project for "the development of the highly advanced computer." Hitachi Ltd. as chief organizer, NEC Corp., Fujitsu Ltd., Toshiba Corp., Mitsubishi Electric Corp., and Oki Electric Industry Co., Ltd. joined this project, which ran 5 years and consumed public funds amounting to around 10 billion yen. Building on the achievement of this project, Hitachi developed HITAC H-8700 (1970) and 8800 (1971), which were installed as mainframes in major universities and national laboratories.

Besides academia and government, these systems also found a variety of uses in industry, such as in seat reservation systems of railway companies and online systems of financial institutions.

As the 1970s progressed, new computers with vector processors appeared. The most famous and successful was the Cray-1, which in 1976 had a peak performance of 160 megaflops. Japanese manufacturers followed suit with their own offerings, including the Fujitsu FACOM VP series, Hitachi S series, and NEC SX series, and the supercomputer business, as it could rightly be called, entered an era of competition between Japan and the United States. In 1981, METI launched a new project called "high-speed computer systems for scientific and technological use." This cooperative project between industry and government had as its objective to develop and operate a 10 gigaflops supercomputer by 1989 and to establish its basic technology in Japan. Although supercomputer technology in Japan was a step short of commercial viability when the project started, Japanese companies produced some of the world's fastest supercomputers, like the Fujitsu VP-100/200, Hitachi HITAC S-810, and NEC SX-1 and 2, soon after the project started. When the project ended in 1989, its initial goal of 10 gigaflops was no longer at the cutting edge of the supercomputer industry. Indeed, Japanese supercomputers reached the highest levels of performance in the mid-1990s thanks to continuous research and development of supercomputers by manufacturers and strong demand from the Japanese manufacturing industry. The Japanese economy was booming and many companies were trying to introduce computer-aided engineering (CAE) in their production processing as a capital investment to cut and shorten development costs and time. At the same time, however, the world-leading supercomputer technology of Hitachi, NEC, and Fujitsu became a source of trade friction with the United States, and together with the subsequent collapse of the economic boom that resonated throughout the 1990s, resulted in some makers' withdrawal from the supercomputer business. METI also changed its policy of supporting the computer industry, leaving it to the National Leadership Supercomputer (NLS) project of the Ministry of Education, Culture, Sports, Science & Technology (MEXT). Figure 7.1 presents a chronology of the world's fastest supercomputers, with the Japanese ones listed on the lower right side of the diagonal trend line.

7.1.1.1 Numerical Wind Tunnel (1993)

The Numerical Wind Tunnel (NWT)[1] was developed as a joint project of the National Aerospace Laboratory, which was once a part of the Science and Technology Agency (STA; now part of MEXT) and Fujitsu. The computer's architecture was one of mixed vector-parallel processors and it was measured at 124 gigaflops on LINPACK in 1993. It remained the world's fastest supercomputer until 1996. Its main application was aeronautical fluid dynamics simulation. The technology of NWT was inherited by the commercial supercomputer, the Fujitsu VPP-500.

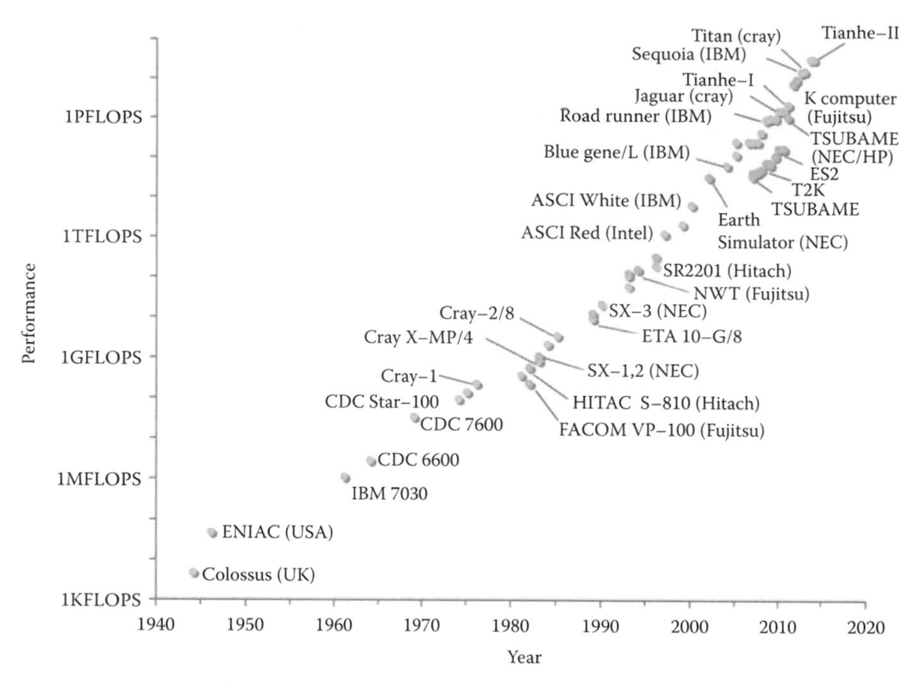

FIGURE 7.1 Trend of the world's fastest supercomputers.

7.1.1.2 CP-PACS (1996)

The CP-PACS[2] project begun in 1991 and was financially supported by MEXT, with Tsukuba University as its leader and Hitachi as manufacturer. In 1996, this massively parallel supercomputer reached a theoretical peak performance of 614 gigaflops with 2048 processors. It went on to make various scientific simulation breakthroughs in particle physics, astrophysics, and condensed matter physics, some of which were based on first-principles calculations. Hitachi produced its SR-2201 commercial supercomputer in 1996 by using the technology acquired through this project.

7.1.1.3 Earth Simulator (2002)

The Earth Simulator (ES)[3] project was initiated in 1998 by STA spending 60 billion yen over 4 years. The system was developed for the Japan Aerospace Exploration Agency (JAXA), Japan Atomic Energy Research Institute (JAERI), and Japan Marine Science and Technology Center (JAMSTEC), with NEC as manufacturer. The system consisted of 640 processor nodes (PN) connected by 640 × 640 single-stage crossbar switches, and each PN consisted of eight vector processors. ES ranked the world's fastest in 2002 (LINPACK performance of 35.86 teraflops). In 2009, it was replaced with a new system based on the commercially available NEC SX-9 (LINPACK performance of 122.4 teraflops). Although the main objectives and initial achievements of ES were to predict global climate changes and geophysics, some of its resources are earmarked for industrial use every year. Accordingly, it has been used for industrial applications based on the fluid dynamics and structural analyses, as well as molecular dynamics and nanomaterials simulations.

7.1.2 Software Development for Industrial Applications and Governmental Support

Compared with its long commitment to hardware development, the history of governmental support for software development in industrial applications has been relatively short. Early in the 1990s when Japanese manufacturing companies were trying to introduce CAE to their production processes, many companies developed their own software. However, after the collapse of Japan's economic bubble in the second half of the 1990s, they ceased to develop software in-house and decided to use commercial software developed in the United States or Europe; these user-friendly packages were well tuned to industrial uses and dramatically reduced the cost and time needed for manufacturing. Feeling a sense of crisis over growing imports of foreign-made software, MEXT began to support development of simulation software for industrial applications by academia in 2002.

7.1.2.1 Frontier Simulation Software for Industrial Science (FY 2002–2005)

In 2002, the Frontier Simulation Software for Industrial Science (FSIS)[4] project was launched as an IT-program research project organized by MEXT and directed by the Institute of Industrial Science (IIS) of the University of Tokyo. The main objective was to develop world-class simulation software for industrial science and to ensure Japan's international competitive position in the following four technological fields: biotechnology, nanotechnology, environment and disaster prevention, and information technology. This project led to systems being developed for quantum chemical simulation, quantum molecular interaction analysis, nanoscale device simulation, fluid dynamics simulation, structural analysis, platform of problem solving environment and HPC middleware. Over 100 researchers from more than 20 institutions participated in this project, and the budget amounted to about 3.8 billion yen over 4 years. Eighteen application software programs were distributed over the Internet since June 2003, and the project was successfully closed in March 2006.

7.1.2.2 Revolutionary Simulation Software Project (FY 2005–2007)

The "Revolutionary simulation software" (RSS21)[5] project succeeded FSIS in 2005 as part of the next-generation IT program sponsored by MEXT and led by IIS of the University of Tokyo. The main aim of this project was to realize revolutionary simulation software that would contribute to rapid progress in science and engineering in the twenty-first century. The project was carried out by the following four research groups: (1) life simulation, (2) multiphysics/scale simulations, (3) safety and environmental simulations of urban areas, and (4) mathematical libraries for HPC. The activities of the last HPC research group included optimization of the software developed in the project for the ES by the Research Organization for Information Science and Technology (RIST). This was the joint project of software and hardware development with an emphasis on industrial applications. Some of the representative software developed through these projects included ABINITY-MP (multiscale simulation of protein and chemical substances for drug design), PHASE (band calculations based on the first-principles pseudopotential method), FrontFlow/red (FFR) (finite volume unstructured CFD for chemically reacting flows), FrontFlow/blue (FFB) (finite element unstructured CFD for aeroacoustics), and FrontSTR (structure analysis). FFR was tuned on the ES, and one of the world's largest Large-Eddy Simulations (LES) of a race car was conducted using 120 million unstructured mesh elements.[6] The scale of the project in terms of funding and total number of researchers involved was almost the same as those of the FSIS project.

7.1.2.3 Research and Development on Innovative Simulation Software (FY 2008–2012)

The 5-year "Research and Development on Innovative Simulation Software" (RISS)[7] project started in 2008. It was supported by MEXT as a program for research and development of next-generation IT infrastructure. It was organized by the Center for Research on Innovative Simulation software (CISS) in IIS of the University of Tokyo. This project inherited much from the FSIS and RSS21 projects. Seven research topics targeted by the project included

1. Thermofluid analysis solvers for large-scale assembly
2. Structural analysis solvers for large-scale assembly
3. Composite material strength and reliability evaluation simulator

4. Large-scale assembly, structural correspondence, multidynamics simulator (these four are in the framework of "next-generation manufacturing simulation systems")
5. Bionanomolecule simulator
6. A biomolecule interaction simulator (these two are in the framework of "quantum biosimulation systems")
7. Quantum function analysis, nanodevice simulator in the framework of "nanodevice simulation systems"

7.2 INDUSTRIAL USERS OF HPC IN JAPAN

7.2.1 Automotive Industry

There is no doubt that the automotive industry is the main user of HPC technology in Japan; in fact, it has been leading the way in HPC simulations for CAE systems for 30 years. Many of the manual design and experimental processes have been digitized and replaced by simulation. It is not going too far to say that new automobiles can no longer be developed without CAE. HPC simulations are extensively used for crash, aerodynamics, and combustion analyses. The most effective simulation in the context of reducing development costs and time is the crash simulation, considering the fact that a prototype vehicle model for experimental crash analysis would cost about US$100,000–200,000. The number of numerical nodes for Finite Element Method (FEM) used in crash analyses has increased by almost a 100 times over 20 years, with the surface resolution at present being 2–5 mm. The rapid development of supercomputers and the lowering of the cost of peak performance have expanded the applications of HPC-CAE in the development process.

7.2.2 Semiconductor Industry

The semiconductor industry uses simulations as indispensable tools for product design. The most popular simulations are the ones for devices, circuits, and manufacturing techniques and are fully mature as technologies. In particular, they are used to design and develop LSI chips. Besides these conventional purposes, the industry is increasing its attention paid to first principles and molecular orbital theory simulations. The main purpose of these new simulations is to predict the solid-state properties of materials. HPC technology is indispensable here too in the context of the need for rapid development and reliability under the strict limitations of

LSI microfabrication processes. The market for nanotechnology in Japan will grow during the next decade, and first-principles simulations based on HPC technology will become more and more important in the LSI industry as a means of studying new materials and devices.

7.2.3 Biomechanical and Biomedical Industry

Among the various biomechanical and biomedical simulations, for example, of blood flow and biomedical tissue deformation, the most promising as far as the industry is concerned is quantum chemical simulations of protein. The most useful application of such molecular dynamics simulations is for designing drugs for the pharmaceutical industry. The main reason for the expensive and ever rising costs of designing new drugs is the low success rate and high failure risk of "seat-of-the-pants" experiments to find a match between the target protein and the chemical compound (drug). Here, computer simulations of chemical bonding based on the quantum chemistry and molecular dynamics may one day replace the need to conduct many experiments and significantly reduce development costs. For precise predictions to be possible, however, molecular dynamics simulations must consider the target protein's dynamic motion amidst water molecules as the medium. In this regard, supercomputers are indispensable. According to Prof. Yasushi Okuno at Kyoto University, the use of the K computer during the design of a new drug can increase the success rate from 1/20,000 to 1/20–1/300 and halve development costs compared to the conventional approach. There is no doubt that drug design using molecular dynamic simulations will be one of the most popular industrial applications of HPC.

7.2.4 Civil Engineering and Construction Industry

Since history began, Japan and its people have been cursed with typhoons, volcanic eruptions, earthquakes, and tsunamis. In addition to the continuous threat of these natural disasters, environmental contaminants have also become an important issue in Japan. In fact, simulation techniques relating to these disasters and environmental assessment are quite advanced in Japan, and the civil engineering and construction industry is a big user of simulations for making disaster and environmental assessments. With regard to the application of HPC, simulations of events such as long-period seismic waves and tsunamis are extremely valuable. Moreover, the availability of HPC resources benefits the construction industry when it comes to making seismic response analyses of structures and wind loading evaluations of tall buildings.

7.3 CURRENT STATUS OF HPC: NATIONAL PROJECTS, INFRASTRUCTURE, AND SERVICES

7.3.1 Overview of the K Computer Project (HPCI Project)

The project titled "Development & Application of Advanced High-Performance Supercomputer" was started in 2006 by MEXT, and reflects the Third Science and Technology Basic Plan set out by the Cabinet Office. The plan itself named supercomputing as an essential technology worthy of support through a national strategy. RIKEN was selected to be the lead developer of a new supercomputer, with the aim of it being the world's fastest, together with its software for nanotechnology and life sciences, and it was also charged with the establishment of a center of excellence for supercomputing research and science. However, after the national elections in 2009, the incoming government reevaluated the project and criticized the way that its development became its own goal and how convincing explanations of its necessity were not provided to the public. After these indictments, the "next-generation supercomputer" project was restructured and the viewpoint of research and development of the supercomputer was changed from the provider side to the user side. Rechristened the "Creation of the Innovative High-Performance Computing Infrastructure (HPCI) project," the main objective became to establish a hierarchical organization; a next-generation supercomputer would be the core, and it would link to other supercomputers at universities, institutions, and a large-scale storage system. The project was to build a consortium of users that would efficiently utilize the HPC infrastructure of Japan. On the basis of this framework, the K computer was built at the RIKEN Advanced Institute for Computational Science (AICS). The K computer broke 8.2 petaflops on LINPACK in June 2011 and became the world's fastest. In November 2011, it broke 10 petaflops. A total of 111 billion yen (US$932 million) was spent on it from FY2006 to FY2012, and the figure includes not only the system development and manufacture (79 billion yen), but also the facilities (19 billion yen) and the software development (13 billion yen) including the optimization for the execution on the K computer.

7.3.2 K Computer and High-Performance Computing Infrastructure

Under the HPCI project led by MEXT, Japanese HPC resources were established and reorganized for public use. The K computer, serving as the National Leadership Supercomputer (NLS), together with other National

Infrastructure Supercomputers (NIS) situated at major universities and institutes all provided their resources to be shared. At present, the common HPC resources amount to a total of about 13–14 petaflops (Figure 7.2), together with the sharing storage of 22 petabytes accessible with a single sign-on system having a common user ID. The HPCI resources are managed in a lump; thus, users (public or otherwise) wishing to use the resources, including the K computer, submit applications all together. The RIST was selected to manage the application process. The selection of applications for public use is based on peer review.

7.3.3 High-Performance Computing Infrastructure Consortium

The HPCI consortium[8] was established in 2012 as part of the activity of the HPCI project. The main purpose of the consortium is, by organizing users of HPCI and supercomputer centers, to unify their opinions regarding the service and operation of HPCI systems, to promote computational science, and to provide suggestions to the government. As of 2013, this consortium consisted of 36 representatives from HPCI user communities, resource providers, and associate members.

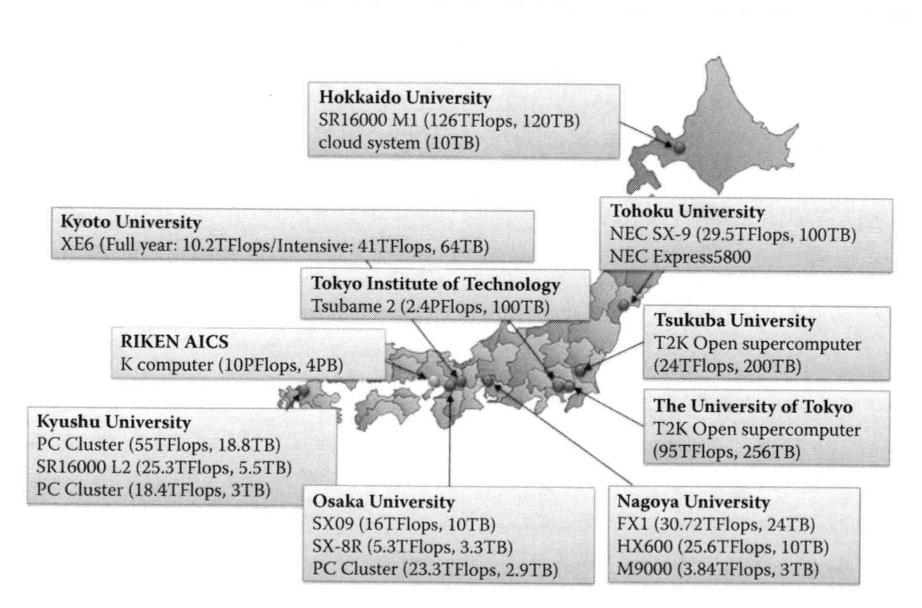

FIGURE 7.2 HPCI Infrastructures and resources provided by each university and institute (total of 13 petaflops for public use).

7.3.4 Strategic Programs for Innovative Research and Field 4 "Industrial Innovation"

During the process of developing K computer, key figures in the HPC field suggested to MEXT that, from social and national viewpoints, the computer should be used strategically in designated scientific areas. This led to the strategic programs for innovative research (SPIRE) project being launched in 2009; five strategic research fields were selected:

1. Supercomputational life science
2. New materials and energy creation
3. Advanced prediction research for natural disaster prevention and reduction
4. Industrial innovation
5. Origin of matter and the universe

The purpose of the project was to establish a framework for promoting and supporting HPC research including human resource development and to produce outstanding social and academic accomplishments by using K computer and other HPCI infrastructures.

After the applications were reviewed, five representative institutes were selected to promote activities in each strategic research field. A feasibility study was carried out (starting in 2010), and a full-scale study lasting 5 years is scheduled to be completed in 2015.

Among the five fields, field 4 is "Industrial innovation,"[9] and IIS of the University of Tokyo is adopted as the representative institution. Field 4 is also being promoted by two other leading institutions, the Japan Atomic Energy Agency (JAEA) and the JAXA. Other affiliated research institutions include Hokkaido University, Tohoku University, Kyoto University, Osaka University, Kobe University, Hiroshima University, Kyushu University, together with 20 more institutions and universities. Thirty companies, including IHI Corp., Hitachi Ltd., Mitsubishi Heavy Industries Ltd., Toyota Motor Corp., Nissan Motor Co., Ltd., Honda R&D Co., Ltd., and Yokohama Rubber Co., Ltd are also participating. The activities include a consortium for studying "vehicle aerodynamics," with 13 companies and 3 universities led by Hokkaido University, a consortium for studying "turbomachinery" with 13 companies and 6 universities and institutions led by the University of Tokyo, and a consortium for studying "combustion and gasification devices" with 6 companies and 3 universities led by Kyoto University.

7.4 RIKEN ADVANCED INSTITUTE FOR COMPUTATIONAL SCIENCE

In 2010, RIKEN established the Advanced Institute for Computational Science (AICS)[10] as the COE of computational science in Japan. The main purpose of AICS is to promote and conduct the cutting-edge research of computational science by tightly linking the hardware and software research and development, through efficient operation of the K computer for users in a vast number of research fields. Besides the operation division to operate the K computer, AICS has a research division consisting of 10 computational science departments, 6 computer science research teams, and 3 research units.

7.4.1 Industrial Committee for Supercomputing Promotion

The Industrial Committee for Supercomputing Promotion (ICSCP)[11] was established in 2005 by the merging of two industrial committees on grid computing research and the FSIS project. Its aim is to promote HPC technology for industrial use, and its activities include evaluation and promotion of "frontier" simulation software based on HPC for industrial applications, implementation of grid computing technology for industrial utilization, enlightenment about supercomputing technology and its introduction to industry, and the development of human resources capable of HPC in industry. As of June 2013, the committee includes 24 regular members together with over 100 registered companies. Their activities are tightly linked with academia and government; lecturers from academia hold seminars on HPC; they disseminate their opinions from an industrial viewpoint and promote national HPC projects on hardware and software.

7.4.2 Distribution of K Computer Resources

In FY2013, half of the K computer's resources were provided to the SPIRE project Strategic Programs, as shown in Figure 7.3. Another 30% was distributed to applications from the public, including 5% to applications submitted from industry. According to its report, RIST selected 69 out of 144 applications for general use, of which industrial use accounted for 35 out of 42 for the use in FY2014. Even though the total amount of resource for industrial applications is relatively low, its higher adoption rate reflects the strong expectation of K computer resources to be used for industrial

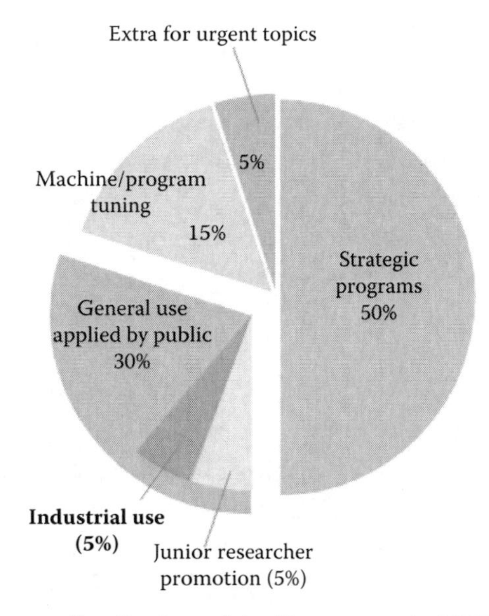

FIGURE 7.3 Resource distribution of the K computer in FY2013.

applications. Besides this general use category, other resources are provided for industrial applications through the SPIRE project. As mentioned earlier, the "Industrial Innovation" field treats research and development of HPC for industrial applications. Through this program, academia and industry conduct joint research that efficiently uses the K computer for industrial applications. The K computer resources distributed to Field 4 in 2013 accounted for slightly more than 10% of the total K computer resource. As a result, about 15% of the total resources were allocated to industrial applications.

Concerning the consortium activities relating to the K computer, the Japan Automobile Manufacturers Association, Inc. made an application to the general use category and together with their own resources used the K computer to conduct a crash safety analysis. In addition, 13 pharmaceutical companies working with 2 universities/institutions organized a consortium on "drug design simulation." The consortium is led by Kyoto University.

7.5 INDUSTRIAL APPLICATIONS ON THE K COMPUTER: ROAD VEHICLE AERODYNAMICS

The HPC Large-Eddy Simulation (LES) of automobile aerodynamics is an example of industrial applications recently conducted on the K computer. The simulation involved making a precise drag prediction of a light motor vehicle using a fully unstructured mesh.[12] A fully automated mesh refinement system was developed for the preprocessing of the FFR developed in the FSIS and RSS21 projects. This work was done by the "Vehicle aerodynamics" consortium of the field 4 in the SPIRE project. In the simulation framework, the user generated about 35 million tetrahedral mesh elements in a conventional CAE environment, and sent this coarse mesh data together with the results obtained with it to the K computer. The preprocessing of the FFR on the K computer involved automatically refining the coarse mesh to the resolution the user wished. The results of the simulation were automatically mapped back to the original coarse grid, so that the user could transfer the results to his/her conventional CAE environment again for visualization without having to deal with a massive amount of data. Figure 7.4 shows the results obtained by a twice-refined mesh in comparison with the original coarse mesh. Compared with the wind-tunnel measurement, the drag of the coarse mesh (35 million mesh elements, surface resolution: several mm) was overestimated by 7%–9%, whereas the twice-refined mesh (2.3 billion mesh elements and submillimeter resolution) had an error of only 1%–2%.

RIKEN AICS has also developed frameworks for other aerodynamics simulations for next-generation vehicle aerodynamics simulations.[13] The main concept of such simulations is to avoid the tedious surface cleanup process when dealing with "dirty" CAD data that includes many gaps

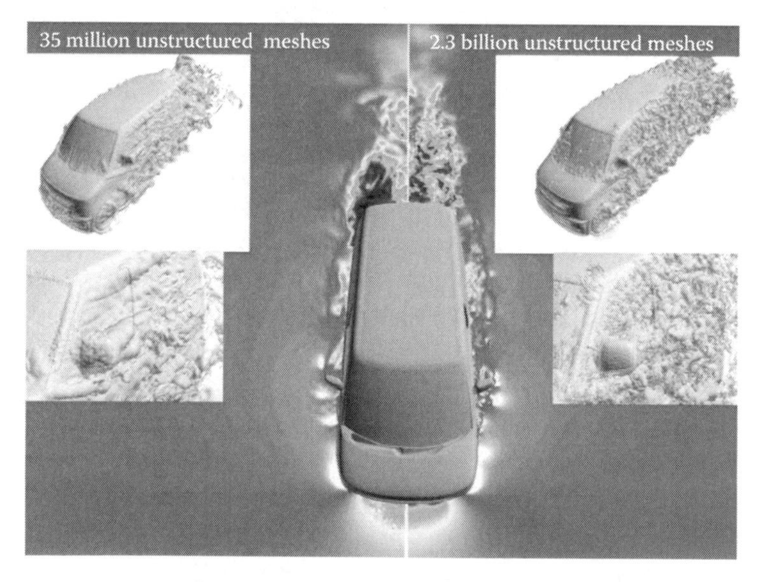

FIGURE 7.4 Large-scale aerodynamics simulation of a road vehicle using fully unstructured 2.3 billion mesh elements (Hokkaido University and Suzuki Motor Corp.).

and overlaps among parts and components. Usually this clean-up process takes weeks to prepare a smooth surface ready for CFD. The automotive industry requires even higher surface resolutions to use CFD to predict aerodynamic forces, and in that context, HPC will be indispensable for some time to come. On the other hand, the cleaning time will increase more if higher surface resolutions are needed; this would seriously impact the applicability of HPC in this industry. To overcome this problem, we have developed a repair-free preprocessing based on a structured grid. With this method, we can generate 19 billion structured mesh elements with a submillimeter surface resolution within a couple of hours from dirty CAD data without having to clean up the surface. Figure 7.5 compares the instantaneous velocity distributions around the vehicle using 110 million (6 mm resolution) and 19 billion (0.8 mm resolution) structured mesh elements. A total of 24,576 nodes (about 200 thousand cores) of the K computer were used to obtain the aerodynamic force data within 1.5 days (including the time taken to generate the mesh).

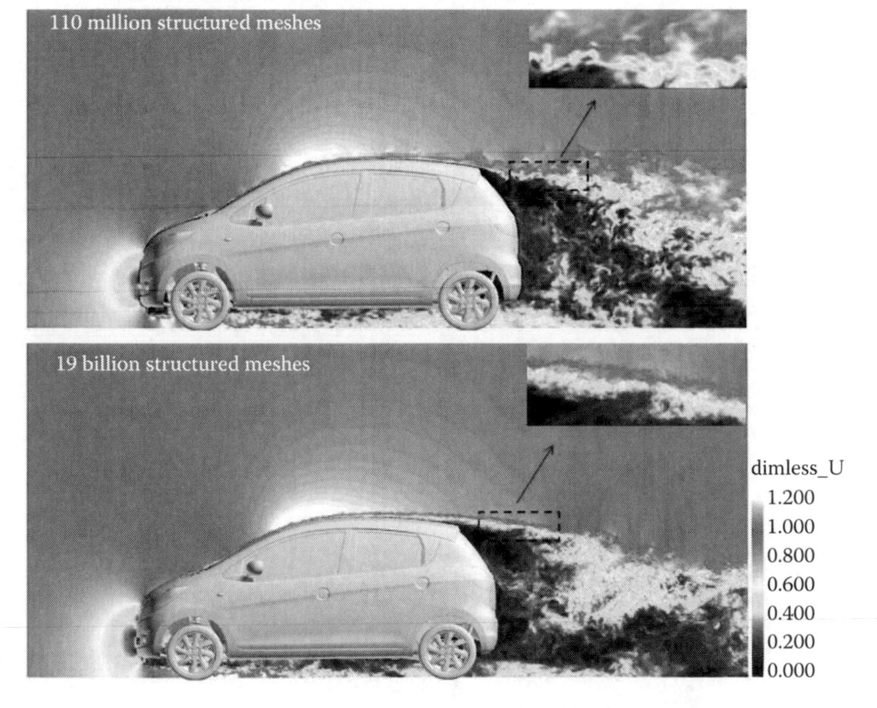

FIGURE 7.5 **(See color insert.)** Next-generation vehicle aerodynamics simulation (RIKEN/AICS and Suzuki Motor Corp.) using 19 billion elements in hierarchical structured meshes.

7.6 SUMMARY

This chapter briefly described the history of HPC in Japan, including hardware and software development and its current status especially regarding industrial applications. At present, MEXT is in charge of HPC policy for Japan, and the K computer serves as the NLS, with other mainframes at universities and national institutions configured around it into an HPCI. The SPIRE project and HPCI consortium are considered to be national and academic support centers for HPC, and industrial applications are promoted within this framework. Japanese industry has been mainly positive about using HPC in their production processes. ICSCP keeps up with their opinions on HPC and promotes HPC policy nationwide. The mainframes of major companies in Japan for CAE use have peak performances of tens to hundreds of teraflops, which is equivalent to the ES 10 years ago. It is expected that the performance of the K computer will be matched by industry within a decade by their own mainframes. Various companies including automobile, heavy industry, and turbomachinery makers want to use the K computer to predict their future CAE needs. Indeed, manufacturing companies installed sophisticated CAE systems for their production process about 10 years ago. It is true that, while some of these systems have not kept up with the rapid progress of both hardware and software in Japan and will soon be out of date, some industrial users are rather negative about replacing their well-thought-out systems. Thus, the application of HPC technology to industrial processes should be discussed not only from the viewpoint of hardware or software performance itself but also in the context of the usability of the total HPC systems including pre- and postprocessing for the treatment of massive data.

REFERENCES

1. T. Iwamiya, T. Nakamura, and M. Yoshida, "Numerical Wind Tunnel (NWT) Project," *Progress of Theoretical Physics Supplement*, No. 122, pp. 57–66, 1996.
2. http://www2.ccs.tsukuba.ac.jp/ccp/cp-pacs.html.
3. http://www.jamstec.go.jp/es/en/index.html.
4. http://www.ciss.iis.u-tokyo.ac.jp/fsis/en/index.html.
5. http://www.ciss.iis.u-tokyo.ac.jp/rss21/en/.
6. M. Tsubokura, K. Kitoh, N. Oshima, T. Nakashima, H. Zhang, K. Onishi and T. Kobayashi, "Large Eddy Simulation of Unsteady Flow around a Formula Car on Earth Simulator," *SAE Journal of Passenger Cars: Mechanical Systems*, 6, pp. 40–49, 2007-01-0106(2007).
7. http://www.ciss.iis.u-tokyo.ac.jp/riss/english/.
8. http://hpci-c.jp/english/index.html.

9. http://www.ciss.iis.u-tokyo.ac.jp/supercomputer/english/.
10. http://www.aics.riken.jp/en/.
11. http://www.icscp.jp/ (*in Japanese*).
12. M. Tsubokura, A. H. Kerr, K. Onishi, and Y. Hashizume, "Vehicle Aerodynamics Simulation for the Next Generation on the K-computer: Part 1 Development of the framework for fully unstructured grids up to 10 billion numerical elements," *SAE International Journal of Passenger Cars: Mechanical* Systems, 7(2): 2014-01-0621(2014).
13. K. Onishi, M. Tsubokura, S. Obayashi, and K. Nakahashi, "Vehicle Aerodynamics Simulation for the Next Generation on the K computer: Part 2 Use of Dirty CAD Data with Modified Cartesian Grid Approach," *SAE International Journal of Passenger Cars: Mechanical Systems*, 7(2): 2014-01-0580(2014).

CHAPTER 8

Industrial Applications of High-Performance Computing in The Netherlands and Belgium

Anwar Osseyran, Peter Michielse, Serge Bogaerts, and Dane Skow

CONTENTS

8.1 THE NETHERLANDS

In the Netherlands, the era of supercomputing started in the early 1980s. Since the introduction of the Cray-1 in 1976 and its apparent success, worldwide interest by academia and large companies grew quickly. Shell,[1] a large Dutch company, was among the early adopters in the Netherlands.

Industrial usage of supercomputing in the Netherlands can be roughly divided into three periods:

1. 1980s and early 1990s: Large industrial companies like Shell and Philips Electronics[2] with vector computers, medium-sized companies with mini vector computers
2. Late 1990s and early 2000s: Transition to workstations and early parallel architectures, that is, parallel computer systems with commodity processors. Usage by smaller companies, but still of significant size
3. Late 2000s until now: Use of scalable parallel architectures with commodity components, scalable architectures. Range of companies becomes larger, with extension to the small- and medium-enterprises (SME) segment

8.1.1 The 1980s and Early 1990s

Shell is one of the largest oil companies in the world, and is involved in both the exploration and production of oil. Apart from that, oil-based products like fuel are produced and sold to customers. The exploration of oil basically entails finding oil. During many decades, so-called seismic processing techniques have been developed to explore the interior of the earth, to predict areas at land or at sea, which may likely contain oil. By sending sound waves into the earth a 3D diagram of the earth layers is constructed, to get an idea of oil-rich earth layers. More sound waves generate more reflections, requiring computer analysis to create accurate diagrams, and as a result, decisions can be made on where to drill and where not. Since drills are very expensive, it is essential to drill right. Avoiding just one unsuccessful drill justifies the investment in computer equipment and in computational methods and algorithms.

The next step is the production of the oil: where exactly to drill, how deep, and how to produce the oil as efficiently as possible (Figure 8.1). Computer simulation of the oil reservoir, and flow of the oil in the earth layers, can be simulated even more accurately to produce oil. Both seismic processing and reservoir simulation were originally done on vector computers in the 1980s but parallel computers would soon

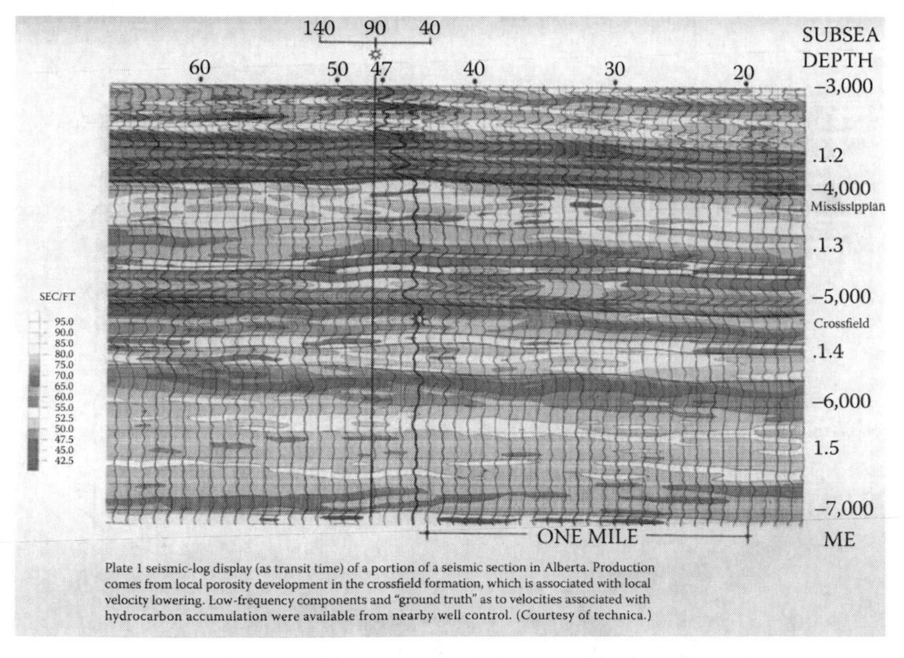

FIGURE 8.1 Visualization of various earth layers in the hunt for oil.

turn out to be more suited for seismic processing and reservoir simulation techniques. Shell experimented with many of these early parallel computing systems.

Quite a few companies followed the example set by Shell: Philips (electronics) modeled high-definition television and chip design and Gasunie[3] (distributor of natural gas in the Netherlands) simulated its gas distribution network on large high-performance computing (HPC) systems, to name a few.

8.1.2 Late 1990s and Early 2000s

Following the use of mini vector computers in the early 1990s, more commodity technology took over. Many companies realized that the price/performance ratio could be improved significantly by using new technology, which ushered in the end of vector systems. An example is the original Organon company, later taken over by Merck. Organon[4] was involved in the research and production of medicine, most notably the contraceptive pill. Organon simulated the interaction between various proteins, based on well-known chemical reactions between atoms. Together with such use, visualization of the results became a key differentiator. Figure 8.2 shows the so-called CAVE, then at SARA,[5] a size of $3 \times 3 \times 3$ m^3, with walls and floor as projecting surfaces. Standing in the CAVE, with special glasses, one could see 3D images and could look inside molecules and try out docking mechanisms.

In this period, companies like Daf Trucks (truck manufacturing[6]), NedCar[7] (car assembly), Solvay[8] (chemistry), GasUnie (gas distribution), Wintershall[9] (oil and gas exploration at the North Sea), Deltares[10] (simulations on water: sea, river, floods, ports, etc.) used HPC systems for their research, production, and visualization needs.

8.1.3 Late 2000s Until Now

In the late 2000s, the third level of companies (small) entered the scene with respect to HPC, either based on using even more commodity technology or relying on access to cloud-based solutions as provided by companies like Amazon.[11]

Interesting examples are small Dutch engineering companies that offer services in the areas of water management, for example, design of ports, simulation of tidal and current flows around dams, and dikes. For a large part, the remainder of this chapter will cover industrial use from this period.

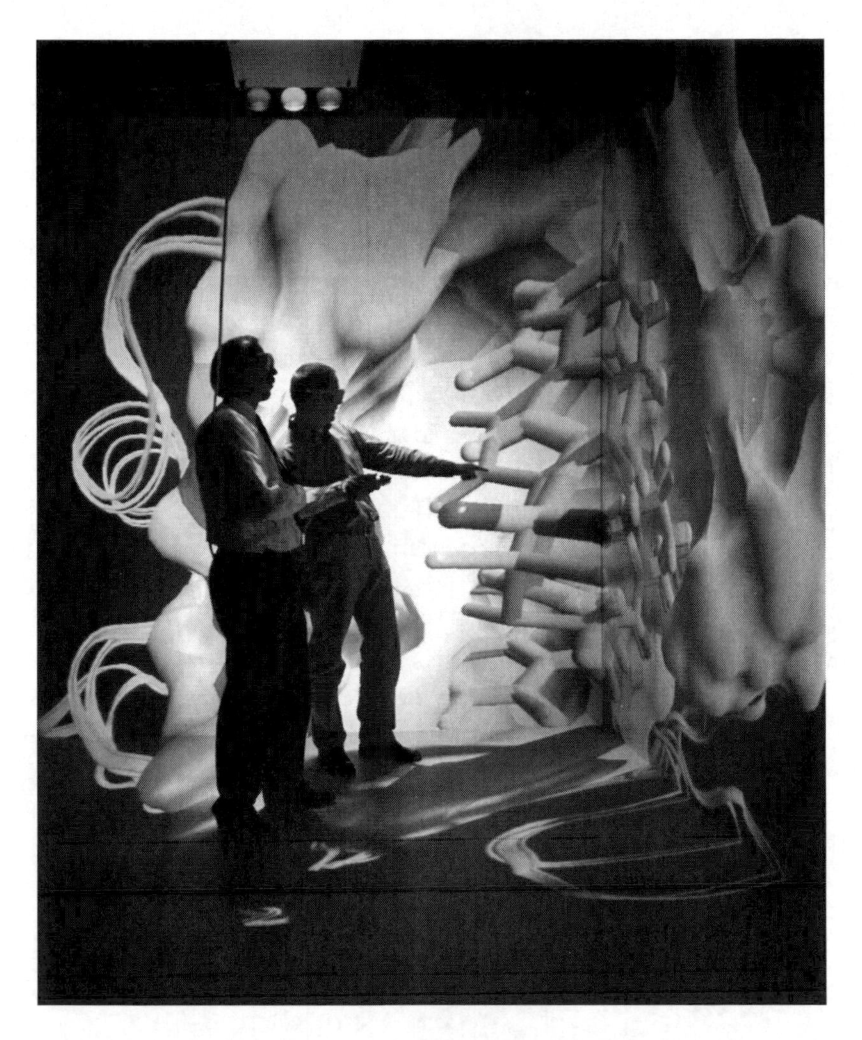

FIGURE 8.2 **(See color insert.)** 3D molecule visualization in the CAVE at SURFsara.

8.1.4 Current HPC Service Provider Landscape

In the early 1990s, the amount of HPC systems in the Netherlands funded with public money grew considerably. Each university with applied science disciplines in its curriculum maintained its own computing center, typically with a small vector computer, later replaced by modest parallel systems. Research centers or institutes, (partially) funded with public money, exploited their own systems. These systems were typically local systems: only meant for usage by researchers at the corresponding institute, and did not have a national user community.

The transition from a large number of local systems to a smaller number of nationally operated supercomputing systems started in the early 2000s. As a result, the current Dutch HPC landscape has evolved over the past years and the HPC availability is currently concentrated in few institutes. Since 2013 SURF, the Dutch collaborative organization for ICT in research and higher education, is responsible for the whole national HPC e-infrastructure. Its subsidiary company SURFsara BV (SURFsara is the Dutch national HPC center supporting research and education in The Netherlands by developing and offering advanced and sustainable HPC infrastructure, services and expertise.) is the national HPC center and SURFnet BV provides the HPC connectivity services. SURFsara offers services in capability computing (national supercomputer services), capacity computing (large compute clusters, HPC clouds and grid), and big data compute facilities and tools, data storage, and visualization services. The national supercomputing service has been available to users since 1984. Its peak performance in floating-point operations per second (FLOPS) has grown from around 100 megaflops (which is 10^8) to the expected 2014 level of more than 1 petaflops (which is equal to 10^{15}). That means a factor of growth of 10 million, over 30 years. Figure 8.3 shows part of the Dutch national supercomputer, named Cartesius.

FIGURE 8.3 (Part of) The current Dutch national supercomputer from Bull, named Cartesius.

During 30 years of service, systems of different HPC vendors have been deployed: Control Data, Cray, Silicon Graphics, IBM, and currently Bull, with the Cartesius system.

The user community in the Netherlands being rather diverse, all national systems have been of general purpose nature. It does not make a lot of sense to install a system meant for national usage if it is well suited to only a certain scientific discipline. Nonetheless, small parts of a system can be equipped with specific hardware to serve certain communities. Examples are large-memory and special processors, such as accelerators.

The University of Groningen has installed HPC equipment for the Lofar project,[12] which until recently included a large Tier-2 IBM Blue Gene supercomputer. The University of Groningen also offers regional HPC compute, storage, and visualization services. Since 2005, Blue Gene systems for Lofar have been used mainly for the task of correlating low-frequency signals from the universe. The Lofar instrument consists of tens of thousands of simple receivers, distributed over a large area. This means that signals from the universe are received by the receivers not simultaneously, but slightly spread in time. Combining the received signals from each of the receivers requires a lot of computational power. Figure 8.4 shows the geographical distribution of antennas in Western Europe and the principle of combining received signals.

Another example is a six-cluster wide-area distributed system (DAS-4[13]) that is deployed by six Dutch institutes to provide a common computational infrastructure for researchers working on parallel, distributed, grid and cloud computing, and large-scale multimedia content analysis. Apart from this, the Dutch Meteorological service KNMI has its own production system

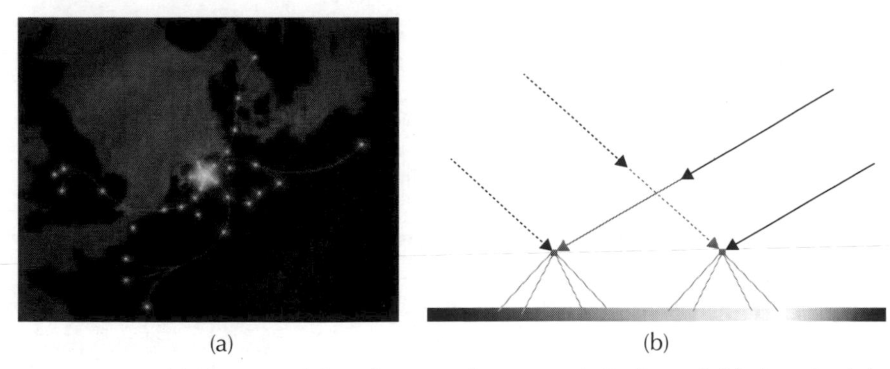

FIGURE 8.4 (a) The actual distribution of antennas in Lofar and (b) the principle of combining received signals.

for weather forecasting.[14] Unlike SURFsara systems, DAS-4 and KNMI systems are not open for researchers developing HPC applications.

8.1.4.1 Dutch HPC Ecosystem

The budget needed for maintaining the Dutch HPC ecosystem is still mainly provided by the Dutch government from education and research funds. Interest from the ministry of economic affairs is also soaring as the competitive benefits of e-infrastructure in general and HPC in particular are becoming more apparent. Other funding sources are provided by the EU through European collaborative projects, research contracts with Dutch industry and technology transfer funds as there is a growing tendency to deploy HPC to increase market penetration of Dutch products and services.

On the industrial side, large Dutch companies fund their own equipment, albeit, in general, large cluster-type systems, that is, good for high throughput computing but less suited for large-capability jobs on many cores. These large companies (such as Shell, Akzo,[15] Unilever,[16] and Philips) usually have their own research facilities in the Netherlands, but companies in the banking and insurance sectors are also increasingly using HPC for portfolio simulations and mathematical insurance modeling. The research collaborations between large companies and Dutch academic research groups are generally well established, leading to indirect access of large industry to the national HPC facilities.

The missing middle between individual HPC users and large corporate HPC applications is very clear in The Netherlands: SMEs rarely have facilities larger than workstation or multicore powerful PCs. There is a lot to be gained here, but this has to be organized. A significant amount of effort will need to be put in both offering hardware and software services and in modeling, simulation, and application development.

8.1.4.2 Dutch HPC Industry

This section describes the Dutch HPC hardware, software, and service providers. Large-scale HPC hardware development and production activities hardly take place in the Netherlands, although companies like ClusterVision play a role as system integrators. ClusterVision[17] specializes in the design, building and management of HPC clusters. The company has grown into one of the leading cluster specialists in Europe. ClusterVision is well known for its management software, which a few years ago led to the spin-off of another company, Bright Computing, which develops and markets the software suite for other cluster vendors as well.

Scientific software development takes place in various areas, ranging from compiler developments to specialized scientific application software. ACE (Associated Computer Experts[18]) is a Dutch company supplying so-called software frameworks to international companies, which in turn develop compiler technology. A compiler is an essential piece of system software, which translates the human-programmed computer code into machine-readable and executable instructions that can be executed in the HPC system. Compilers make use of the hardware features of the processor chip and the overall architecture of the system and are key components in obtaining performance.

An interesting company that deserves some attention is ASML,[19] the world's leading provider of lithography systems for the semiconductor industry. Based in Veldhoven, the Netherlands, ASML develops and manufactures complex machines that are critical to the production of integrated circuits or chips. It is amazing to see that the ASML lithography machines are able to work on a few tens of nanometers size, with 40,000 laser signals per second on a tiny area. This is why clean-room technology is key, or current chips would have huge amounts of disturbances and hence could not work properly.

If investments in technology and development of new lithography systems seem huge, rewards are equally huge. Another example is the mirror technology used in these machines. Mirrors are polished to subnanometer accuracy, so the surface is as flat as possible. Blown up to the size of the Netherlands, the biggest difference in height would be less than a millimeter. This is an amazing accuracy, which is expected to be improved, as ASML foresees still further technological progress in chip manufacturing machines, which are essential to maintain chips to follow Moore's law.

There are many start-up companies in the Netherlands in the area of specific application development, targeted to academia and industry for use on HPC systems. Some of these have emerged from research disciplines at Dutch universities and public research institutes. For example, SCM (Scientific Computing & Modelling) has its origin at the VU University Amsterdam, and develops chemical simulation software for molecular modeling. This software gives insight into chemical bonds in complex molecules, and is an essential tool for industry to simulate behavior and effect of new medicines before synthesis of the actual medicine.

A special example is Deltares, which has emerged from Delft Hydraulics and GeoDelft as an independent institute for applied research in the field of water, subsurface, and infrastructure. Throughout the world, Deltares works on smart solutions, innovations, and applications for people, environment, and society, with its main focus on deltas, coastal regions, and river basins. Development and use of application software in these areas, to be run on large computer systems, is an important asset for Deltares. Examples are in the areas of coastal waters and estuaries, rivers, dike stability, and so on.

In 2011, the Dutch government decided to support and invest in innovation activities of Dutch industry and SMEs, which include the activities and fields mentioned earlier in this section. This initiative is known as the Dutch "topsector" initiative, in which 10 areas of industrial activity, which are crucial for the Dutch economy, have been identified. Areas that directly benefiting from HPC deployment in a broad sense are water management, chemistry, energy, and agriculture, but other sectors such as the creative industries, high-tech systems and materials, health, and logistics may also become main users.

8.1.5 Developing The Dutch Quantum Supercomputer

The Netherlands are set to work on building the next generation of computers, the quantum computer, under supervision by TU Delft. For this purpose, science, industry, and the government have formed the QuTech Institute.[20] QuTech is the result of collaboration in the field of innovation. "The best researchers are working with the most innovative entrepreneurs and with the government to develop a revolutionary technique for creating new products," according to Henk Kamp, Dutch Minister of Economic Affairs. "This can lead to socially relevant new products and services, such as predicting the effect of medicine, and of course to more income and jobs in The Netherlands."

The new computer should be a reality by 2030, and will use the special characteristics of small particles, the so-called quantumbits. These small particles can deliver an enormous computing power, thereby making new applications possible. For example, a quantum computer would be able to calculate the effect of drugs per individual. It will also help calculate and predict the properties of certain materials and soil layers.

8.1.5.1 Dutch Government HPC Policy

The Dutch government HPC policy was reviewed in 2009. Ecorys,[21] a research and consulting company, was contracted by the Dutch government to conduct an investigation into the economic effects of the deployment of supercomputers in the Netherlands (Amsterdam), Switzerland

(Lugano), Germany (Jülich), Spain (Barcelona), and the United States (San Diego). Their report[22] resulted in a formal statement summarizing the vision of the government on supercomputing in the Netherlands.[23] The main conclusion is that major future societal challenges in the Netherlands require a strong commitment to the deployment of HPC for scenario development and analysis. The physical presence of HPC infrastructure and facilities is required for solving major social problems, promoting key areas in computational science, and strengthening economic structures through innovation and creating jobs. The government recognized that supercomputer infrastructure and services have positive societal and economic impact and lead to competitive advantages in various industrial fields as well as help train less skilled personnel. The Dutch government commits to play a major role in the European Strategy Forum on Research Infrastructures (ESFRI) and in the European Tier-0 project PaRtnership for Advanced Computing in Europe (PRACE).

8.1.6 Case Studies And Business Cases

8.1.6.1 Case Study: The Oosterschelde Storm Surge Barrier

The Oosterschelde Storm Surge Barrier in the southwest of the Netherlands is generally recognized as one of the modern seven wonders of the world. The storm surge barrier was the most difficult part to build as well as the most expensive part of the so-called Delta Works, in place since 1950 to protect large parts of the low Netherlands against storms and high sea levels. The infamous flood of 1953,[24] which killed almost 2000 people and flooded about 1500 km^2 of land, proved beyond doubt the importance and relevance of the Delta Works.

The initial plan was to shut by means of dams all large river mouths of Rhine, Meuse, and Schelde, except those essential for the economically important ports of Rotterdam and Antwerp (Belgium). After many public and governmental discussions, it was decided in the early 1970s to not fully close the biggest river mouth (the Eastern Schelde) by means of a dam, but to build an open barrier containing a number of sluices that would only be closed during heavy storms and high water levels. This way, the unique freshwater environment and the favorable fishery conditions in the Eastern Schelde would be maintained. Sixty-two openings, each 40 m wide, would be installed to allow as much salt water through as possible, while maintaining the tidal movement. The Eastern Schelde storm surge barrier turned out to be one of the biggest structures of the world.

The costs of this "open dam" were considerably higher than the costs of an ordinary closed dam: €2.5 billion was needed to complete the barrier.

One of the key challenges to its construction was the required stability of the piers between which the movable sluices would be constructed. The tidal effect in the Eastern Schelde is large, and combined with high water levels and heavy storms from the west, the water force onto the piers would be extremely large. Therefore, an accurate simulation of the water flow conditions both with and without piers was absolutely essential to determine the right dimensions of the piers: too heavy would be unnecessarily costly, too light would be disastrous in case of extreme weather and tidal conditions. Figure 8.5 shows the Oosterschelde Storm Surge Barrier in open mode (left, in calm conditions) and closed (right, in extreme conditions).

Under the leadership of the Dutch Ministry of Water Works, which was responsible for the project, a consortium of mainly Dutch construction companies modeled the actual water flow conditions both on a scale model and with computer simulations. It turned out to be relatively easy to predict water levels, but predicting water flow speeds proved more difficult. The main difficulty was the fact that the tidal effects are so large that the sand bottom of the Eastern Schelde is in continuous motion, which affects water flow speeds. This gave rise to the development of new numerical simulation techniques and computer codes, executed on large Amdahl and Control Data supercomputing systems. The combination of better techniques and large systems is what HPC is about: in this particular Oosterschelde case, the required dimensions (weight, size, height, and placement on the sea bottom) of the piers could be determined accurately, thereby avoiding huge costs and future disasters.

FIGURE 8.5 The Oosterschelde storm surge barrier.

8.1.6.2 Case Study: Optimization of Steel Flow in the Mold of a Continuous Caster

Tata Steel[25,26] in IJmuiden, formerly called Hoogovens, serves customers who demand high-quality steel. Production of such steel quality requires optimum production processes, which in turn are being simulated and developed as a result of multidisciplinary research on material properties and the underlying fluid dynamics of the casting process. In the casting process, liquid steel is poured into a mold from above. Around the outside, heat escapes from the steel creating a layer of solidified steel, which is extracted from the bottom of the mold. Magnetic fields ensure the correct flow of steel through the mold. This casting process is called "continuous casting" (see Figure 8.6).

For Tata Steel, it is important to use cost-effective solutions for modeling and simulation of steel flow in the mold of a continuous caster to meet the demand for a faster, more complex, and better-engineered manufacturing process. The physics covered in the computational fluid dynamics (CFD) model cover the simulation of incompressible mass, momentum

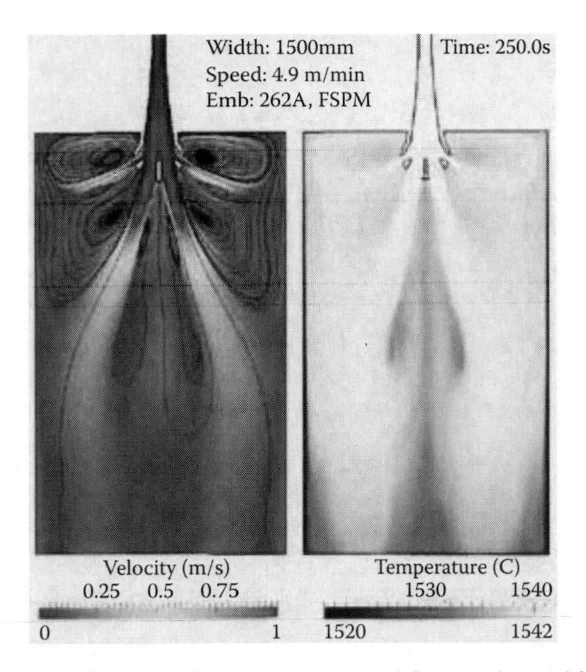

FIGURE 8.6 **(See color insert.)** Temperature and flow in the mold at the Direct Sheet Plant (DSP). This specialized thin casting machine conducts the solidified steel from the mold directly to the rollers, and can reach very high casting speeds. Good quality numerical simulations are required to understand the flows.

and enthalpy equations, and a solidification model with magnetohydrodynamics. The costs of the necessary HPC infrastructure are almost negligible compared to the financial waste caused by any malfunction during the production process. As steelmaking is a bulk industry, minor improvements to the process deliver major cost savings and make the business case of HPC deployment a no-brainer.

An interesting aspect is the magnetic field that is used to "guide" the flowing steel. Little is known about the effects of magnetic fields. Areas of research cover the strength of the magnet, and the orientation of poles. An interesting side effect is the use of open-source software instead of commercial (and much more expensive) application software. The fact that open-source software can be used further improves the business case for using HPC equipment and applications to optimally model the casting process.

8.1.6.3 Case Study: HPC in the Financial Sector[27]

Most people are familiar with pictures of the hectic trading of shares on Wall Street or other stock market exchanges: traders are running around, shouting and gesticulating, apparently busy with making money. The exact right or wrong moment of selling or buying stock can mean millions of dollars of profit, or loss. It is no surprise then that financial institutions use automated trading policies to determine the exact right moment to sell or buy.

There are, however, many more areas in the financial industry where computer simulations are used to assess as accurately as possible the consequences of financial transactions and events that may happen on the financial markets. For instance, financial institutions rely heavily on numerical modeling and computation for the quantification of the fair-value and risk inherent to trading in financial derivatives contracts. A well-known method for estimating risks is based on historical data used to generate possible future scenarios. This example is trivial in terms of computation but when one considers portfolios consisting of thousands of instruments, including exotic options that need to be revalued many times, the workload soon grows substantial.

As a response to the severe losses encountered during the recent credit crisis, regulators have recently introduced several new requirements that financial institutions need to fulfill to strengthen their capital and liquidity positions. The so-called Basel III regulatory framework includes important new risk and valuation measures. Together, these new requirements introduce a severe burden to the modeling and computation framework from the perspective of modeling complexity, data quality, and computational speed. As a result, financial institutions like banks, insurance companies, and pension and hedge funds are increasingly applying HPC in their day-to-day business. This ranges from so-called Monte-Carlo methods (in which many estimates are done based on simulated "market behavior") to traditional finite difference methods, which are used to solve partial differential equations describing the fair risk premium as a function of the underlying market factors.

It is interesting to note that not only traditional HPC architectures but also more special-purpose hardware is being used in these "conservative" environments. At ING Bank, both architectures are adopted: a grid of multicore CPUs and a grid of GPUs depending on the specific application and the numerical solver that is used for the corresponding calculations. The increased computational requirements as a result of the new regulatory landscape are not solely responsible for adopting HPC: changing markets have also significantly increased the complexity of valuation and risk models. Valuation and risk models for products previously known as simple, or "vanilla," have become much more complex mainly due to the different new hybrid risk components that the market is actively pricing after the recent credit and liquidity crisis. A trade-off between a CPU or GPU or even a hybrid HPC technology solution is predominantly driven by the numerical approaches that are in place. Monte-Carlo based methods are highly efficient and cost–effective on GPU systems, whereas models based on partial differential equations favor CPU-based parallel systems more.

Another important development that is taking place is the shift toward in-house development. Although in previous years many calculations were done using external vendor systems, the general strategy now is to move toward internal solutions where the modeling, software development, and HPC technology choice are under full control of the firm.[27] This shows that the actual investments to control everything in-house are a very clear indication of the Return-on-Investment for financial institutions [1,2].

8.1.7 Stimulation Plans for Adoption of HPC by The Dutch Industry and SMEs

As stated earlier, the Dutch academic and public research communities rely currently on the national HPC facilities at SURFsara. The gap between available systems, if any, in industry and SMEs on one side and

SURFsara on the other side is generally large. The Dutch government has therefore launched a plan to focus on stimulating the access to the large national HPC facilities. This plan, with some legal constraints (avoiding unfair competition), has been developed and is now being enrolled in the Netherlands. The plan focuses not only on early adopters or large multinationals but mainly on potential new application areas. The focus is also on incorporating HPC in the whole value chain of existing HPC users.

8.1.8 Extension To Big Data

HPC and Big Data technologies and services are increasingly merging. Use of high-end computing leads to new science, and as a result to new market opportunities for commercial companies. Dutch SMEs working in the life sciences area, for instance, have to cope with an ever-increasing amount of data. Access to facilities to analyze large datasets produced by simulations but also by DNA sequencers and other sources has become crucial. HPC services should therefore incorporate a Big Data offering in their catalogues. Various Dutch Big Data applications are using the HPC services of SURFsara. A technology transfer program between SURFsara and its daughter Vancis[28] offering commercial ICT services has also been developed to stimulate the use of HPC by Dutch SMEs.

SURFsara supports large-scale data analytics with its Apache Hadoop service. The service makes fast data-parallel analysis of very large datasets possible by providing interfaces that enable very fast experimentation on such data, without having to bother with the specifics of parallelism. Combined with the commodity hardware on which Hadoop runs, it is one of the most efficient data-parallel computing platforms currently available. After a year of testing on a prototype facility at SARA, the Hadoop production facility was made available for scientists in the Netherlands by the end of 2011. The service supports scientists in the domains of Information Retrieval, Natural Language Processing, Computer Science, and Bioinformatics in advancing their research, making use of very large amounts of data that were not fully exploited so far. The production facility complements the Dutch national computing infrastructure as a system for data-parallel, high-throughput capacity computing. Because of its simplicity we believe the service will be an important enabler of large-scale computing and e-Science for researchers that are not traditionally involved in these fields.

An interesting example is the growing interest from companies, governments, and universities in the daily communication that takes place on online social media such as blogs, Facebook, and Twitter. Linguists and researchers in communication studies can use this data to study language variation and change. Companies may track reputation of a product after its introduction. Journalists may follow the spread of news messages and spot initial local reports of incidents. Police may monitor Twitter for suspicious behavior (Figure 8.7). However, the amount of social media data is large and obtaining specific parts that are interesting for a certain purpose, is not easy.

Together with partners,[29] the Netherlands e-Science Center has initiated a so-called Path Finding Project, which aims at developing a centralized service for gathering, storing, and analyzing Twitter messages and making available derived information to a consortium of researchers in communication studies and language technology throughout the Netherlands. The service will be based on an existing systems set up at the Universities of Amsterdam and Groningen with infrastructure from SURFsara. The Twitter API, providing free access to approximately 1% of all tweets worldwide, is constantly harvested and the resulting data stored. Interfaces to this data provide users with a number of analysis tools that can be run on all content and metadata.

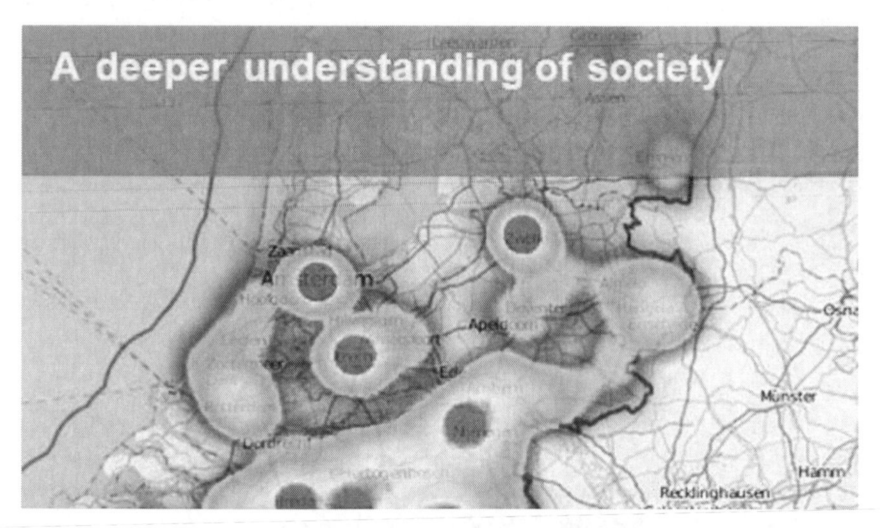

FIGURE 8.7 **(See color insert.)** Occurrence of word "carnaval" in tweets. Carnaval[30] is an event celebrated primarily in the southern provinces of The Netherlands. The results show where Carnaval is a hot topic on tweets, and enables retail stores to adapt their products in this period, for police to concentrate manpower in the right places, etc.

8.2 BELGIUM

8.2.1 Overview of HPC Governance in Belgium

Governance of HPC in Belgium is distributed among its Federated Entities (Regions and Communities) in accordance with their respective competencies. The regional HPC infrastructure in Flanders is provided through grants from the Hercules Foundation (Herculesstichting[31]) (The Hercules Foundation is a government agency supporting fundamental and basic strategic research for all scientific disciplines in Flanders. This may range from databases and collections to telescopes and supercomputers. At the time of writing the manuscript, Dane Skow was HPC director at the Herculesstichting.). The mandate to Hercules from the Flemish government is to provide HPC resources to academic researchers in the Flemish universities, the Flemish research centers, and to Flemish industry (on a cost recovery basis). Tier-2 facilities are provided at each of the Universities and a regional Tier-1 facility is provided at one of them, the Ghent University (UGent). Funding of HPC facilities and staff for the French-speaking universities, academia of the *Fédération Wallonie-Bruxelles*, is mainly provided by the *Fonds de la Recherche Scientifique—FNRS*,[32] and coordinated by the *Consortium des Équipements de Calcul Intensif –CÉCI.*[33] This consortium gathers the HPC centers of five of the French-speaking Universities and makes sure investments in Tier-2 systems of these institutions are optimized. Moreover, Wallonia is funding HPC through regional research programs supporting economy and targeting innovation: applied research centers are privileged beneficiaries of these programs and, among these centers, Cenaero[34] is focusing on numberical simulation and HPC. (Cenaero performs technology transfer and provides in-depth CAE expertise to industries in the Wallonian region of Belgium. In addition to that, Cenaero is involved in R&D-projects with industry and academia.). The Walloon Tier-1 system is funded by Wallonia, who appointed Cenaero as operator of the facility that is shared between academic partners, applied research centers, and industry. Coordination of HPC initiatives at the Federal level, which includes the involvement of Belgium with international programs such as PRACE, is settled by the *Interministerial Conference for Science Policy.* Wallonia was later appointed for organization of the representation of Belgium in PRACE and created a specialized coordination group, the *CIS/INFRA/PRACE*, bringing together representatives of the above-mentioned actors of HPC in Belgium. The Belgian Science Policy Office is funding the membership fee of Belgium in the PRACE organization.

8.2.2 HPC Resources

The Flemish Tier-1 system (Muk) is a 175 teraflop machine with dedicated 10Gbps network links to all five Flemish universities. Most industrial research using the Flemish HPC infrastructure is done in partnership with one of the Flemish universities. Typically, this is research done by a joint team of university and industrial researchers. We are seeing early requests for direct access to the HPC resources and these are also currently routed through the University connections.

The Walloon Tier-1 supercomputer is a heterogeneous machine of 11496 cores. The CÉCI gathers in each of the five associated Universities Tier-2 systems with complementary characteristics. They range from large SMP node machines to clusters of several thousands of cores. These systems are accessible to all researchers of the associated Universities and are used in the frame of both fundamental and applied research projects, including collaborative projects with industries. Similarly, Cenaero leverages HPC resources for applied and industrial research through collaborative research projects as well as advanced engineering services for commercial industrial needs.

8.2.3 Exemplary Industrial HPC Users

8.2.3.1 Techspace Aero (Safran) Designs Efficient Low Pressure Compressors (Wallonia)

Techspace Aero (Liège, Belgium),[35] a Safran company,[36] which is a leading international high-technology group active in aerospace, defense, and security, is producing low-pressure compressors for several modern jet engines. This activity includes the design of the corresponding engine module that relies significantly on numerical simulation techniques. The engineering department of Techspace Aero has for several years been collaborating with Cenaero in developing advanced simulation methodologies, for example, in the framework of European or regional research programs, that allow them to enhance their designs. These advanced methodologies allow to improve the prediction of complex physical phenomena or to simulate details of the engine design that impact, for example, the engine stability in certain operation regime. The implementation of these methodologies was performed by Cenaero through the use of high-performance clusters and parallel codes. Figure 8.8 shows the actual design of the compressor module (left) and the simulated pressure distribution in the compressor (right).

When Techspace Aero considered integrating these methodologies in their standard engineering practices, they naturally considered Cenaero as the HPC resources provider aware of their technical needs and able to cope with their industrial constraints (reliability and security). The implemented solution is integrated in their design environment provided by the group. Techspace Aero engineers prepare their simulations with their usual tool from which they can choose to submit their computations to several sites including Cenaero (i.e., with resource provider diversification and hence security). This integration provides transparency in data transfer and remote job execution for the user.

The use of numerical simulation allows Techspace Aero to offer competitive engine components targeting 15% consumption decrease compared to standard designs, thereby maintaining its competitive edge on the market.

8.2.3.2 3E Forecasting Optimal Wind Turbine Sites (Flanders)

3E[37] is a global consultancy and software company focused on increasing renewable energy installation performance and optimizing energy consumption. In 2013, they approached UGent to get access to their Tier-2 facilities with a view to developing a wind energy resource estimation consulting tool built around the open-source Weather Research

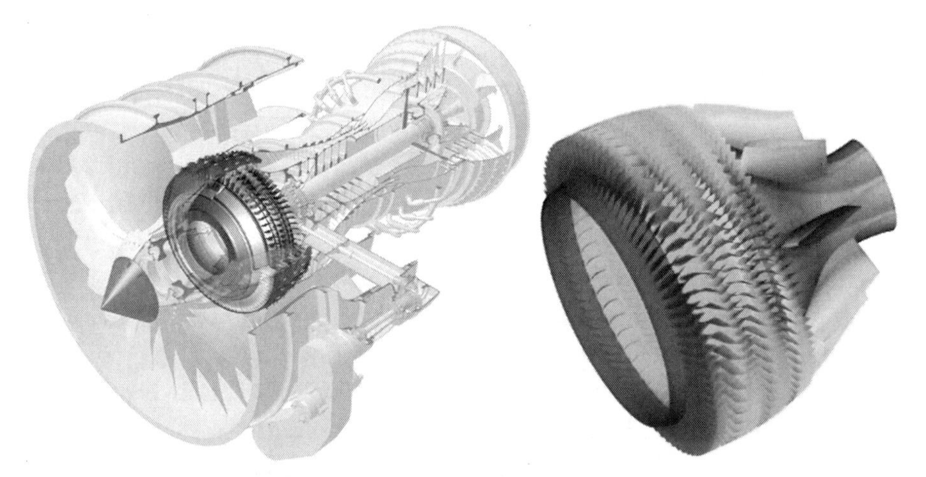

FIGURE 8.8 **(See color insert.)** Low-pressure compressor module in a jet engine and example of simulation result. (courtesy of Techspace Aero [safran].)

and Forecasting (WRF) software. Initial tests were done with a 10-km geographic grid, yet even so, estimations based on long-term simulations (typically 20 years) required computational resources beyond those available to their research and development team. This caused 3E to seek HPC resources outside the company.

Contact was made with the HPC team at UGent, which has significant experience managing large, computationally intensive, open-source software on HPC systems, and a collaboration begun. The initial tests were reworked and refined in consultation between the 3E engineers, developers, and the UGent team. The resulting package ran much faster on the Tier-2 systems and enabled the 3E team to pursue an order of magnitude deduction in grid size—enabling much finer resolution of wind features in the simulated environment. However, since the computational work grew as the cube of the number of grid cells increased, this would require 729 times as much computing power—much more than was practical even on the UGent Tier-2 systems. Enter the Flemish Tier-1 system.

With early access to Muk, 3E was able to take the original package, reduce the grid cell size, and complete qualification runs within 3 months at the new resolution of 1 km. Figure 8.9 below qualitatively shows the improvements in the resulting model forecasts, enabling them to find new viable placements for profitable wind turbines and to more accurately predict long-term energy production and return on investments for

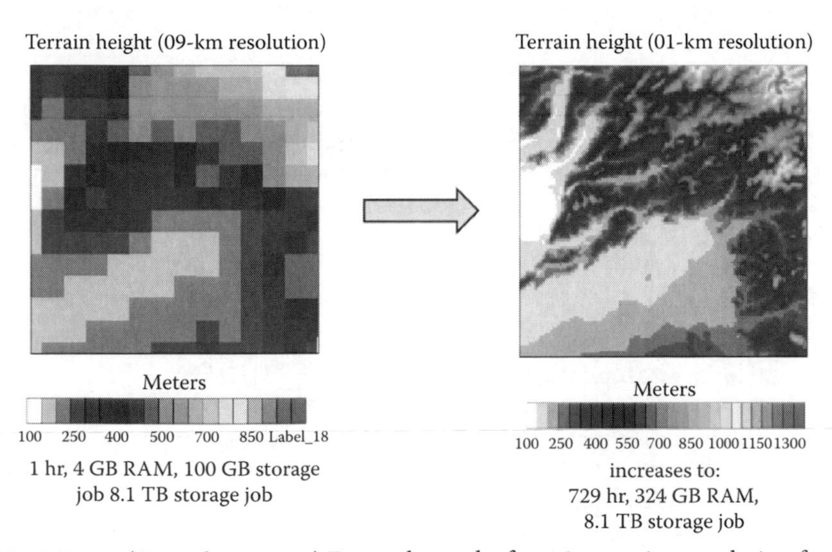

FIGURE 8.9 **(See color insert.)** Example results from improving resolution from 9 to 1 km.

developers. 3E is now in the process of turning this package into a production consulting tool for use by their customer engineering teams. A new line of product was born within a few months by having access to 1000× the computational power they, as a medium-sized business, could have afforded in-house [3,4,5].

8.3 CONCLUSION

The Netherlands, and increasingly Belgium as well, have been able to maintain HPC infrastructure that is suited to its variety of users from research and industry. Not only the expected scientific (beta) disciplines but also alpha and gamma sciences find their way to using the available HPC infrastructure. The challenge for the future is clearly twofold. First is to maintain and further improve the current level of HPC services, covering supercomputing, big data analysis, data storage and data management, support, and domain expertise. Second is to continue and extend the involvement of SMEs and industry to HPC resources, and to find more ways to attract SMEs and larger industrial companies to use the available resources and support.

REFERENCES

1. J. C. Hull, 2006 Options, Futures and Other Derivatives (Sixth Edition), Upper Saddle River, NJ: Prentice Hall International.
2. BCBS, 2010 Basel III: A global regulatory framework for more resilient banks and banking systems, December 2010.
3. 3E Company: www.3e.be.
4. R. Donnelly, "HPC for Industrial Wind Energy Applications," (3E), VSC Users Day 2014 presentation, http://www.vscentrum.be/en/Morton et al., "Pushing WRF to its Computational Limits," AWS10, http://weather.arsc.edu/Events/AWS10/Presentations/MortonBenchmark.pdf.
5. WRF software: http://www.wrf-model.org/index.php.

ENDNOTES

1. http://www.shell.com.
2. http:www.philips.com.
3. http://www.gasunie.nl.
4. http://www.organon.com.
5. SARA is the national computing and data center in The Netherlands. It changed its name to SURFsara in 2013.
6. http://www.daf.com.
7. http://www.vdlnedcar.nl.
8. http://www.solvay.com.
9. http://www.wintershall.com.

10. http://www.deltares.nl.
11. http://www.amazon.com.
12. http://www.lofar.org/node/38.
13. http://www.cs.vu.nl/das4/.
14. http://www.knmi.nl/cms/content/105587/nieuwe_knmi-computer_operationeel.
15. http://www.akzo.com.
16. http://www.unilever.com.
17. http://www.clustervision.com.
18. http://www.ace.nl.
19. http://www.asml.com.
20. http://www.qutech.com.
21. http://www.ecorys.com.
22. https://zoek.officielebekendmakingen.nl/blg-48238.pdf.
23. http://www.tweedekamer.nl/downloads/document/index.jsp?id=4a1127e7-8e56-4032-984f-320ddc3372cf&title=Kabinetsvisie%20supercomputers%20en%20supernode%20in%20Nederland.doc.
24. http://en.wikipedia.org/wiki/Flood_of_1953.
25. Dr. Eelco van Vliet, Tata Steel, IJmuiden, The Netherlands.
26. http://www.tatasteel.nl.
27. Dr. Drona Kandhai, Quantitative Analytics, ING Bank, Bijlmerdreef 98, 1102 CT Amsterdam, The Netherlands.
28. http://www.vancis.nl.
29. See more at: http://www.esciencecenter.nl/projects/path-finding-projects/a-deeper-understanding-of-society/#sthash.yr8wru36.dpuf. Partners: Netherlands e-Science Center, University of Groningen, Radboud University Nijmegen, University of Twente, University of Amsterdam, SURFsara, Erasmus University Rotterdam.
30. http://en.wikipedia.org/wiki/Carnival_in_The_Netherlands.
31. http://www.herculesstichting.be.
32. http://www.fnrs.be.
33. http://www.ceci-hpc.be.
34. http://www.cenaero.be.
35. http://www.techspace-aero.com.
36. http://www.safran-group.com.
37. http://www.3ecompany.com.

CHAPTER 9

Industrial High-Performance Computing Activities in Korea

Sang Min Lee

CONTENTS

9.1 A BRIEF HISTORY ON THE KISTI NATIONAL INSTITUTE OF SUPERCOMPUTING AND NETWORKING

The Korea Institute of Science and Technology Information (KISTI) National Institute of Supercomputing and Networking (NISN) began in a computer room of the Korea Institute of Science and Technology (KIST) in June 1967. Since KIST's establishment in February 1966, this computer room evolved into a central institute leading the Korean information society, with computing technology magnified as one of its major research areas. The KIST computer room was reformed as the System Engineering Research Institute (SERI) in November 1984 and started in earnest to carry out supercomputing operations in 1988. SERI played a big part in establishing the validity of centralizing national computing and research network resources in a national center by introducing and operating the first supercomputing system in South Korea in 1988.

This institute pulled the supercomputing application ability up to the level of leading countries and contributed to open global research networks, helping South Korea become a leading country in the information industry in the twenty-first century. South Korea started its supercomputing operations with the introduction of the CRAY-2S system in 1988. The CRAY-2S was actively used in various research fields and in many industrial areas. CRAY-2S reached its maximum capacity as a result of an increasing number of users and workloads, creating a strong desire to introduce a new supercomputing system. As a result, in 1993, Korea's second supercomputer, the CRAY-YMP C90, was installed.

With the introduction of the CRAY-YMP C90 in 1994, the second phase of the Korean supercomputer operation program began. The first phase of the program (from 1988 to 1993) entailed fostering the user environment of the supercomputer and activating the use of the supercomputer, whereas the second phase focused on research and technology development related to supercomputing and elevated supercomputing applications to the level of leading countries. To maximize the use of the supercomputer, user groups in terms of specific research domains were organized. These domain groups were encouraged to contribute by sharing their supercomputing technology and research products.

The ratio of economic outcome to investment was reported to be around 38:1 in the period of the Korean supercomputing program from 1998 to 2000. Korean government sufficiently invested in the supercomputing program, and the active support from Korean government made the program larger and stronger. KISTI encouraged supercomputer users to efficiently utilize the supercomputer and increased internal researchers' competence to extend the supercomputing basis. From 2001 to 2003, the third Korean supercomputing system, IBM p690 and NEC SX series, was successfully introduced. From 2009 to present, KISTI has operated the fourth Korean supercomputing system, which consisted of a large SUN Linux cluster and IBM p595 systems (Table 9.1).

KISTI could hereby provide supercomputing resources to Korean computational scientists and industrial R&D people. With the promotion of

TABLE 9.1 History of KISTI Supercomputers

Generation	1	2	—	3	4
Operation period [year]	1988–1993	1993–1998	1997–2001 CRAY T3E [115GF]	2001–2009 IBM p690 [4 TF]	2009–present Sun Linux [324 TF]
System name [performance]	CRAY 2S [2GF]	CRAY C90 [16GF]	HP GS320 [111 GF]	NEC SX-5/6 [240 GF]	IBM p595 [37 TF]

the supercomputing program, a scientific research networking project (Korea Research Networking Program, hereafter KREONET) was satisfactorily progressed. The project has continued to establish a world top-level research network. In 2013, KREONET covered the whole Korean peninsula with 14 regional networking subcenters.

In 2000, SERI was integrated into KISTI and the Korean supercomputing program was taken over by the KISTI supercomputing center (KSC). In 2013, KSC was reformed into KISTI NISN as a result of the Korea High-Performance Computing Act 2012.

9.2 KOREA SME STATUS AND STRATEGIC POSITIONING FOR SME SUPPORT

In all ages and countries, industry plays an important role in the development of a nation's economy. Therefore, it is imperative for the government of nations to pursue independent encouragement and support policies. Many companies, particularly small- and medium-sized enterprises (SMEs), are having a difficult time in the midst of the fierce competition in the current free-world market. Some are said to be considering closing their operations and laying off their employees to remain competitive with higher labor overhead, ascendant production costs, and old-fashioned technology.

To overcome the crisis, industry needs to take a revolutionary step and look for technologies that can empower them to make products of better quality, reduce production costs, and quickly analyze designs for the parts and whole assembly of their product. Product design engineering is going through big changes with modern digital technologies. The development of technology along with the progress of computing power provides companies with the opportunity to attain higher productivity in geometrical product modeling, assembling, and instruction manuals. The computing technology advances along with software engineering, and then plays a key role in tiding over the crisis.

Computer modeling and simulation has brought about dramatic innovation in engineering as well as in scientific areas. Based on computing technology, traditional engineering has evolved into a new research and practical area, namely CAE (computer-aided engineering) that can give companies innovative progress in manufacturing products of high quality and low cost by replacing experimental physical methods with simulation methods when testing their products. Today, high-performance computers—that is, supercomputers—are used as the main technical development equipment in industry.

To gain market share, companies need to have the best product at the lowest cost. Supercomputing in industry is often associated with product design that can control the quality of the whole product manufacturing process. Product design by supercomputing is often achieved in the same way as general engineering product design is with personal computers. Contrary to CAE with personal computers, the main benefit of using supercomputers is a drastic reduction of time and cost related to development and manufacturing of products. These modern technologies give users of so-called industrial supercomputing a competitive advantage.

Even companies that fully understand the effectiveness and availability of supercomputing are still confronted with impediments for industrial use of supercomputers. The impediments usually stem from two causes: the nature of industry itself and aspects related to required support. In this chapter, we discuss the causes of impedance to industrial supercomputing and we propose a supercomputing service model for SMEs to attain technology innovation through industrial supercomputing. And finally we present some success stories obtained by small companies through industrial supercomputing.

Most companies, especially small ones, have barriers to use supercomputing infrastructure that generally are defined as supercomputer access and application software in their R&D processes. The reasons why they cannot actively use supercomputing infrastructure originate from two causes: limitations within the companies themselves and broader limitations affecting entire industry sectors.

Generally, SMEs do not have sufficient manpower for R&D or the internal organization to make an appropriate product design on account of their weak capital, and are realizing the importance and utility of the new technology in their R&D. Nevertheless, they still find themselves seized by a vague fear for changing from traditional methods to new modern technology. And there is really insufficient basic internal infrastructure for internal developers or product designers to follow the new supercomputing technology.

To extricate many industries from the current vulnerable and inefficient situation, external support efforts are definitely necessary, which is why most of the parties concerned have provided support policies related to industrial supercomputing, public infrastructure for industrial R&D, and industrial supercomputing technical support. If a cooperative system for invigorating industrial supercomputing was established among SMEs, universities, and governmental institutions; SMEs could reach the technical innovation faster and easier.

Traditionally, supercomputing infrastructure is mostly used in big science fields; for example, evolution of the universe, exploration of new substances and natural resources, nuclear weapon experiments, and strategic simulations. Many major companies have been well supplied with HPC infrastructure and high-quality human resources, thereby causing affluent growth by the benefit of a powerful support organization, supercomputers, and CAE tools. In the present day, it is very important and inevitable for SMEs to optimize a given supercomputing resource especially in the process of product design. Because they often have insufficient manpower and R&D infrastructure, it is not easy for SMEs to use supercomputing R&D resources as readily as large companies and major scientists do. Most of the SMEs have this practical gap in using supercomputers. Governmental institutes, which lead supercomputing technology and can modify the technologies into ones suitable for use by SMEs, should set up an appropriate service model for industrial supercomputing. Only then can SMEs board the train of technology innovation.

To increase industrial competitiveness, it is necessary for industry to conduct product design with high precision and low cost by using supercomputing technology. But most companies do not invest in supercomputing infrastructure and commercial application software, and specialists in engineering supercomputing are lacking in most companies. Recognizing these shortages, KISTI inaugurated an industrial technology support project for 78 SMEs through the high-speed application support program from 1998 to 2000. And then, in 2003, KISTI selected three SMEs in thermodynamics and structural fields, and supported them in developing product designs under tight collaboration with industry, academia, and institutions with supercomputing technology. Since 2004 the project was implemented as a nation-wide program.

In 2007, KISTI started the SME supercomputing program under the budget support of the Korean government's Small and Medium-size Business Administration (SMBA). The program proposed to facilitate Korean SMEs like global enterprises. KISTI domain specialists have supported SME's R&D people in the core steps of product development, that is, product design and evaluation steps, with engineering design, visualization, and CAE simulation based on supercomputing technology, so that cost and time for product R&D are dramatically reduced up to about 45%.

KISTI has provided technical support to SMEs from 2007 to present as follows: forty-nine SMEs in 2007 with an R&D budget of $5 million, about forty SMEs per year from 2008 to present with an R&D budget of $5–6 million. A book containing success stories of R&D through the use of supercomputing technology has been published every year.

9.3 AN INDUSTRIAL SUPERCOMPUTING SERVICE MODEL

Supercomputing is not only about technology but also about the people and organizations who are key to the further progress of these technologies and about the complex web that connects people, organizations, products, and technologies. From the point of view of supercomputing, the infrastructure for supercomputing must be closely integrated with five layers: (1) fundamental hardware infrastructure (supercomputer hardware itself, high speed networks, data storage system, visualization system, etc.), (2) operation software infrastructure (operating system, job management, resource management, etc.), (3) application infrastructure (application software, graphical user interface, etc.), (4) a sustainable system for the technology transfer among users, academia, institutions and governmental organization, and finally (5) purpose-oriented usage of the supercomputing infrastructure. Veritable supercomputing can be realized only when the layers mentioned above are satisfied to optimum levels. Sadly, this is not the case in all of industry today.

Traditional industrial developers have come to understand that using computing tools at the high end of the performance spectrum within various areas of disciplines can provide a competitive opportunity in product design quality. Nevertheless, small companies have not really engaged themselves in the computing scene. Moreover, scarcity of R&D man power and regional differences in the employment situation are prevalent in industry. So, for small industries such as SMEs to benefit from supercomputing, it is necessary to establish a stable and sustainable supercomputing service model.

Basically, a supercomputing service model for small industry should be designed with the basic concept of supercomputing, based on the reality and needs of the industry users. Keeping in mind the direction, we designed an industrial supercomputing service model, so-called SME Supercomputing. The model is composed of four components: (1) supercomputing R&D environment targeted to small firms, (2) joint research and consulting aimed at small companies, (3) engineering education and training, and (4) operation by a coordinative committee formed from industry, academia, and institutions.

9.3.1 Supercomputing R&D Environment for SME

It is very difficult for SMEs to use supercomputing infrastructure by themselves. Actually, they have to obtain the optimized product design from supercomputer and engineering software. Thus, the supercomputing R&D environment for SMEs should be set up, keeping in mind the actualities of small companies. KISTI has established the actual demand and present situation of SMEs, and has built two kinds of product design environments. Two high-performance computing (HPC) R&D environments for SMEs are provided through a web portal.

Generally, SME developers utilize R&D to create and improve the products through several design processes including validation steps. In the processes, SMEs are very interested to know how the product design could reflect consumer's demand. That is, an optimized cycle of product design and validation would be very beneficial to get to the right products in a short amount of time. To enhance the SME's design and validation process, KISTI has combined product design and validation processes with CAE software under the supercomputing environment to build an automatic product design environment that is operated on a web portal. On the portal, SME developers can get a report of optimized product design. KISTI has established the "Fan Simulator" (e.g., see Figure 9.1) as the first outcome with the concept of "adaptive computing" based on HPC environment.

As for the general product design process, the automatic product design environment surely has its limitations; for instance, it is impossible to customize the design portal for each and every product. This means that another product design environment is needed for SME developers to make a suitable and optimized product design platform covering geometric design, product CAE simulation, and optimization processes through CAE technology based on HPC. KISTI has established a "large-scale realistic design platform" (e.g., see Figure 9.2) for the structural mechanics domain as the first outcome of the "virtual product design environment" project.

FIGURE 9.1 An automatic product design environment: The fan simulator can treat to design "Axial" and "Sirroco" fan types.

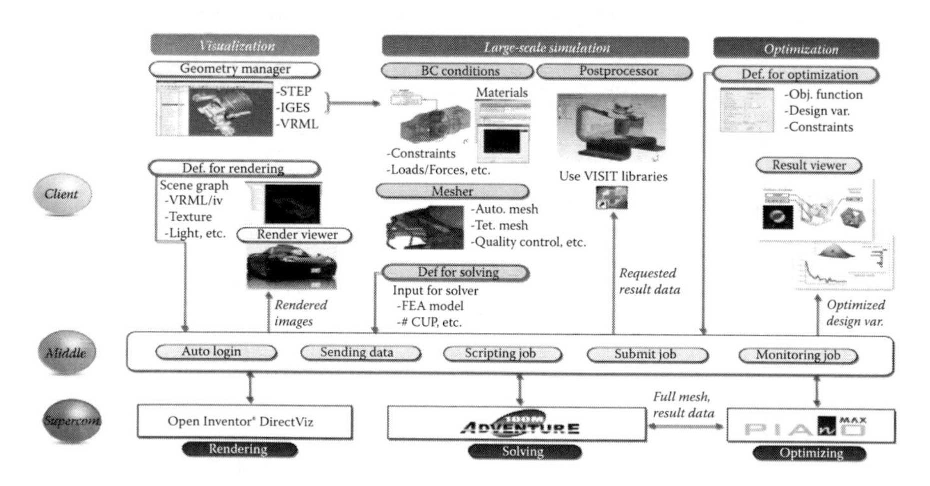

FIGURE 9.2 Architecture of virtual product design environment.

9.3.2 Joint Research and Consulting

SMEs are often under pressure to meet the deadline of a new product launch. Small companies put forward a proposal to a governmental industry R&D program; for example, the National SME supercomputing R&D program, which then performs a rigorous screening process. When the proposal is honored, the project is pushed forward with collaboration between an SME developer and a domain specialist at KISTI supercomputing center who together set up an R&D roadmap for the project. The specialist provides engineering and supercomputing techniques to the SME partner and support for product design, CAE simulation, and a review with the SME developer's demands. In the process of the joint research, the specialist frequently gives valuable advice with regard to obtaining an engineering model and choosing CAE software (Table 9.2).

The light-colored bars in Figure 9.3 denote the number of joint research projects in various engineering domains selected through a rigorous screening process for the National SME supercomputing R&D program from 2007 to 2013. The average selection rate is around 25%. Korean government funds 75% of the individual project R&D investments to the selected project proposers (SME+KISTI). The selected SME invests the remaining 25% of the project's R&D investment (5% in cash and 20% in kind). Usually, a joint SME project runs for 1 to 2 years, depending on the difficulty and the scale of the industrial product problem.

The dark bars in Figure 9.3 reflect the number of joint projects that were selected through a screening process of KISTI's SME supercomputing engineering support program. KISTI fully supports the SME with R&D investment and CAE technology based on HPC from KISTI. This joint project runs approximately 4 or 5 months.

TABLE 9.2 Status of SME Supercomputing Project from 2007 to 2013

Years	2007	2008	2009	2010	2011	2012	2013	Total
No. of SME supported	49	44	30	15	38	61	31	268
R&D Fund [$MUSD total per SME]	4.9/0.1	6.6/0.15	6/0.2	3/0.2	3.8/0.1	14.75/0.25	7.5/0.25	46.55/0.18

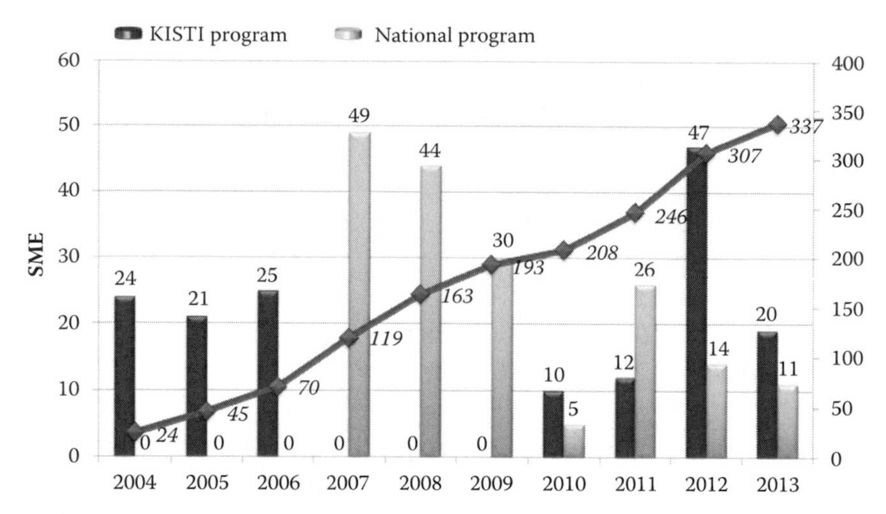

FIGURE 9.3 Annual change of SME Supercomputing project. Light bars delineate the number of SME supported by Korea government agency, SMBA and dark ones KISTI itself. The upward-sloping line reflects cumulative annual number of SME supercomputing subjects.

9.3.3 Engineering Education and Training

Engineering education and training is needed for SME developers to maintain their R&D without any assistance from specialists. In the education program, software specialists teach usage methods of the supercomputer for commercial engineering software (e.g., CFX, FLUENT, ANSYS, ABAQUS, and LS-DYNA). The education typically includes training with realistic problems.

To foster the younger students with strong engineering skills based on supercomputing; KISTI—along with regional universities and local SMEs—carries the "Supercomputing Digital Manufacturing Track." In the track, that opens for about a month in both summer and winter vacation, CAE specialists give students and SME developers practical lectures on CAE modeling, simulation, and analysis for product design. After completing the course, students receive credits and vocational qualifications. KISTI then recommends candidates to industries for employment.

9.3.4 Operation of Coordinative Committee

Generally, whenever companies plan to release a new product, they take the strategic approach for dominating the market in advance. However, small companies do not have sufficient means for laying out a complete plan for a new business. To assist SMEs when launching a plan for a product, we have initiated a coordinative committee composed of specialists skilled at CAE engineering, analysis of market response, and supercomputing. Committee members will typically belong to various organizations: institutions, universities, or industry. They are consulted for the market prospects of a new product, master-plan of CAE simulation, and revaluation of products.

9.4 SUCCESS STORIES

9.4.1 Usage Effects of Supercomputing

The SME supercomputing service model has been applied to Korean small industries since 2009. We have performed the outcome analysis of the SME supercomputing service by a survey for 91 industries. The survey mainly focused on investigating the usage effects of the supercomputing service: cost and time reductions of product development before and after using supercomputing.

According to the survey, the average costs to develop a product ran to about $456,000 before using supercomputers. When supercomputing technology was used, the costs reduced to about $211,000 (reduction ratio about 53.4%). Similarly, the average developing time is about 3755 hours per product. Using supercomputers, this reduces to about 1787 hours (reduction ratio about 52.4%).

9.4.2 Representative Industry Outcomes

Small companies selected for the National SME Supercomputing Support Program are generally supported by specialists with skills of CAE and supercomputing who belong to the KISTI supercomputing center. Actually, every problem of the selected companies can be solved with the use of supercomputer. We present some success stories for SME Supercomputing service as follows.

In 2007, electric automobile company CT&T, planned to manufacture a new lightweight car with an inner air-conditioning system. We established a joint research team between CT&T and the KISTI supercomputing center, and pushed ahead with the R&D plan over the course of 2 years. The electric car, "E-ZONE," met the requirements for collision safety, heating and conditioning, and had a defrosting system at the windshield. We used ABAQUS for the collision safety problem and FLUENT for the heating/air-conditioning and defrost system simulation. KISTI supercomputer

GAIA (IBM p6) was used for the car simulations with 16 and 24 cores of CPUs during about 3000 and 5000 hours respectively (e.g., see Figure 9.4). It would normally cost CT&T about $3 million to design a new car. Through using HPC technology, development cost was reduced to about $300,000 (reduction ratio of 90%). CT&T were contracted by the California Police Agency to manufacture 2,000 E-ZONEs for parking attendants and in Korea approximately 40,000 E-ZONEs were sold by contract.

We have supported the company Kyungan Jeonsun to develop a no-pilot aircraft for a missile target. Key issue for the joint research was to verify whether a primary design for the aircraft satisfied some strict qualifications: the craft should fly over 500 km/h without the body shaking. We performed a flight simulation on the same supercomputer with FLUENT using 16 cores of CPUs during about 4000 hours (e.g., see Figure 9.5).

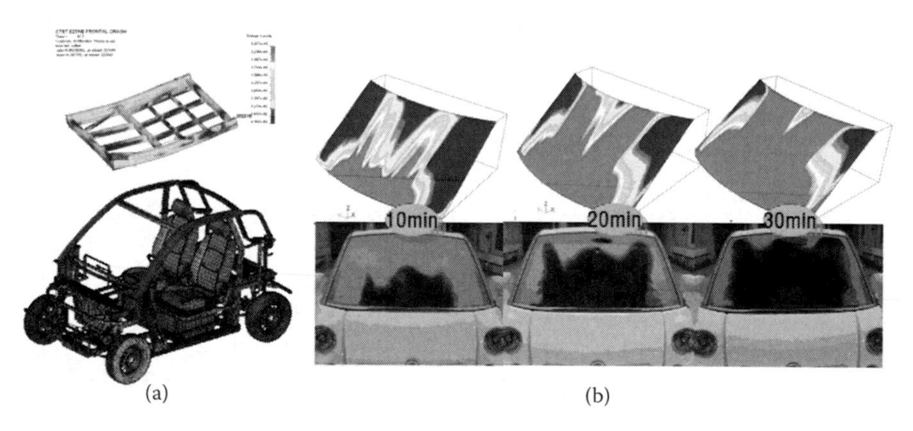

FIGURE 9.4 **(See color insert.)** (a) Structural Analysis (b) Defrost Simulation of e-ZONE.

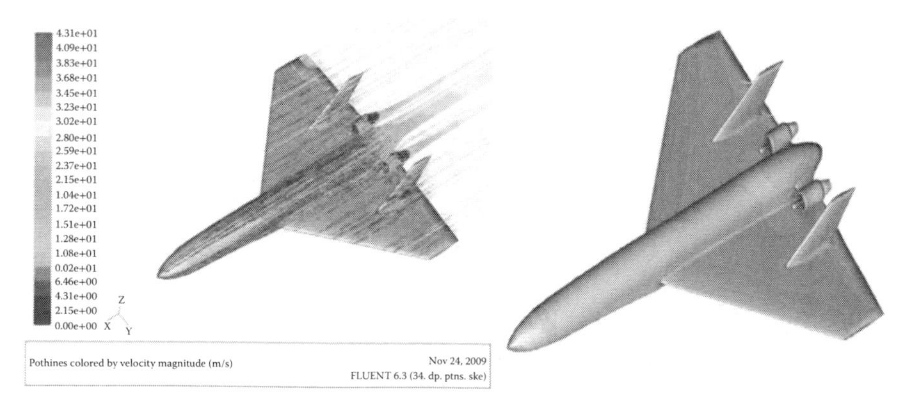

FIGURE 9.5 **(See color insert.)** Pilotless aircraft simulation.

It would normally cost Kyungan Jeonsun approximately $1 million to design a no-pilot aircraft. Through using HPC technology, development cost was reduced to about $100,000 (reduction ratio 90%). Kyungan Jeonsun was contracted by the Korean Army Headquarters to provide pilotless aircraft for missile targets.

9.5 CONCLUSION

We have proposed a supercomputing service model for small industry. Through the model, SMEs have the opportunity to take their technological innovation to a new level. We can confirm that using supercomputing gives small industries high economic effects with a time and cost reduction ratio of over 50%. Consequently, the Korea SME supercomputing service model has played a prominent role in contributing to the advancement of small industry.

CHAPTER 10

Industrial Applications on Supercomputing in Spain

Francesc Subirada

CONTENTS

10.1 INTRODUCTION

Are supercomputers a necessity or a luxury in Spain?

The effective and efficient management of public resources should always be paramount in our society, especially in times of economic crisis such as the one we are undergoing today. It is imperative to use all the decreasing resources at our disposal to develop activities that improve the life quality of the citizens and also create wealth for everyone. It is logical that society asks itself whether it is convenient to invest dozens of millions of euros in research, development, and high technology in moments like these, with all sorts of difficulties to overcome.

So, the question is indeed pertinent: are supercomputers a necessity or a luxury in Spain?

Let us take a look at history.

10.2 HISTORY

It is important to clarify that this chapter does not intend to offer an exhaustive explanation of the situation of high-performance computing (HPC) in Spain. We will be focusing on its fundamental player, the Barcelona Supercomputing Center (BSC-CNS), appointed as a National Facility by the Spanish government in 2005. The first HPC systems and the pioneering industrial usage of HPC systems in Spain date from the late 1980s to early 1990s, covering the use of vector computing, followed by various parallel computing systems. A number of public regional centers initiated their service at that time.

The BSC-CNS inherited the tradition of the well-known European Center for Parallelism of Barcelona (CEPBA). This institution was a research, development, and innovation center on efficient computing technologies both for academia and for industry. CEPBA belonged to the Technical University of Catalonia (UPC).

CEPBA started its activities in 1991, gathering the experience and needs from different UPC departments. The Computer Architecture Department (DAC) provided experience in the lower level of computing systems (numerical kernels, operating systems, tools, and architecture). Five other departments of the UPC with high computation demands joined DAC to set up the CEPBA. These departments were Signal Theory and Communications, Strength of Materials and Structural Engineering, Computer Systems and Languages, Nuclear Physics and Engineering, and Applied Physics.

From 1995 to 2000 CEPBA coordinated the service activities with CESCA (Supercomputing Center of Catalonia) through the C4 (Computing and Communications Center of Catalonia). In 2000, CEPBA signed an agreement with IBM to launch the CEPBA-IBM Research Institute (CIRI). The objectives of this agreement were joint research on topics related with deep computing and architecture, and supporting local research in other areas of science and engineering. This Research and Development Partnership between UPC and IBM had an initial commitment for 4 years.

In 2004, the Ministry of Education (Spanish government), the Generalitat de Catalunya (local Catalan government) and the UPC took the initiative to create a National Supercomputing Center in Barcelona, the BSC-CNS, as a continuation of the CIRI.

So, the MareNostrum system was built in 2004 as the most powerful machine in Europe and the fourth most powerful supercomputer in the world. In 2005, the BSC-CNS was officially constituted and started its activities. When upgraded in 2006 shortly after its establishment, the MareNostrum system was the most powerful supercomputer in Europe again and the fifth most powerful supercomputer in the world (Figure 10.1).

Because it considered supercomputing infrastructure and services to be a decisive asset for the scientific and technological development of the country, the Spanish government created the Spanish Supercomputing Network (Red Española de Supercomputación, RES) in 2006 answering to the need of the Spanish scientific community for intensive calculation resources.

The RES consists of a distributed virtual infrastructure of supercomputers located in different sites and each contributes to the total processing power available to users of different R&D groups in Spain. Its operation is coordinated by the Operations Department of the BSC-CNS, which includes support for global maintenance and upgrades, training of users and technicians, facilitation of access, and other aspects related to user support.

From 2007 to 2013, the BSC-CNS led the Spanish national Consolider project "Supercomputación y e-Ciencia," which brings together 21 of the

FIGURE 10.1 **(See Color Insert.)** MareNostrum's home in Barcelona's Chapel Torre Girona.

best Spanish research teams to collaborate in designing supercomputing applications to efficiently use HPC architectures.

In 2012, MareNostrum underwent another major upgrade, which has allowed it to reach 1.1 petaflops of performance. MareNostrum III is one of only six nodes in four countries that together form the pan-European PRACE (Partnership for Advanced Computing in Europe) Tier-0 network, which provides world-class supercomputing services to European scientists, and as such 70% of its capacity is dedicated to PRACE.

10.3 CURRENT HPC SERVICE PROVIDER LANDSCAPE AND ECOSYSTEM

The current Spanish HPC landscape has been evolving over the past years and the HPC services offered are currently concentrated in the following centers:

- Agencia Estatal de Meteorología—AEMET: http://www.aemet.es/es/portada
- Barcelona Supercomputing Center—Centro Nacional de Supercomputación (BSC-CNS): www.bsc.es
- CénitS—Centro Extremeño de Investigación, Innovación Tecnológica y Supercomputación—LUSITANIA: http://www.computaex.es/cenits
- Centre Nacional d'Anàlisi Genòmica (CNAG)—http://www.cnag.cat/
- Centro de Investigaciones Energéticas, Medioambientales y Tecnológicas—CIEMAT—http://www.ciemat.es/
- Centro de Supercomputación de Galicia (CESGA)—http://www.cesga.es/
- FCSCL—Fundación Centro de Supercomputación en Castilla y León: http://www.fcsc.es/index.php/es/
- Instituto de Astrofísica de Canarias (IAC)*—LaPalma Node: http://www.iac.es/
- Instituto Tecnológico de Canarias (ITC)*—Atlante Node: http://www.itccanarias.org/

- Parque Científico Murcia—centro de Supercomputación—Ben Arabi: http://www.parquecientificomurcia.es/web/centro-de-supercomputacion/general
- Servicio Santander de Supercomputación (SSC)*—Universidad de Cantabria—Altamira Node: http://www.unican.es/
- Universidad de Cádiz: http://www.uca.es/
- Universidad de Granada—Alhambra: http://csirc.ugr.es/informatica/supercomputacion/
- Universidad de Málaga*—Pablo Picasso Node: http://www.uma.es/
- Universidad Politécnica de Madrid (UPM)*—Magerit Node: http://www.upm.es/institucional
- Universitat de València*—Tirant Node: http://www.uv.es/
- Universidad de Zaragoza*—CaesarAgusta Node: http://www.unizar.es/

Those centers marked with an asterisk above are members of the RES, which is a virtual distributed infrastructure at the service of supercomputing R&D in Spain. It entails the interconnection of a group of free associated supercomputers that share their assigned tasks, and are managed in a joint and efficient way, thus maximizing their computing power.

The aim of the RES is to provide world-class HPC resources and services to meet the ever-increasing demand of the Spanish and international scientific communities. Despite the augmentation of available resources, the amount of requests exceeds available capacity.

The RES also organizes training sessions for both technicians managing the infrastructure, and scientists as well as seminars to enhance the dissemination of scientific results. Since the creation of the RES in 2006, more than 2050 projects have been awarded access to the network, illustrating the relevance and need of access to supercomputing facilities in Spain (Figure 10.2).

European HPC-facilities are organized in a pyramid according to the volume of computing resources offered and the number of systems that provide them. The shared European vision is to encourage and support the creation of an overall European HPC ecosystem involving all stakeholders: HPC service providers, grid infrastructures, scientific and industrial user communities, and the European HPC hardware and software industry.

The RES, as a national and local-level HPC service provider, intends to provide Tier-1 and Tier-2 level infrastructures. The upgraded MareNostrum III, as a PRACE infrastructure, provides Tier-0 service to Europe and service at Tier-1 level to Spain via the RES network. The other RES nodes will provide Tier-1 or Tier-2 services according to their capabilities (Figure 10.3).

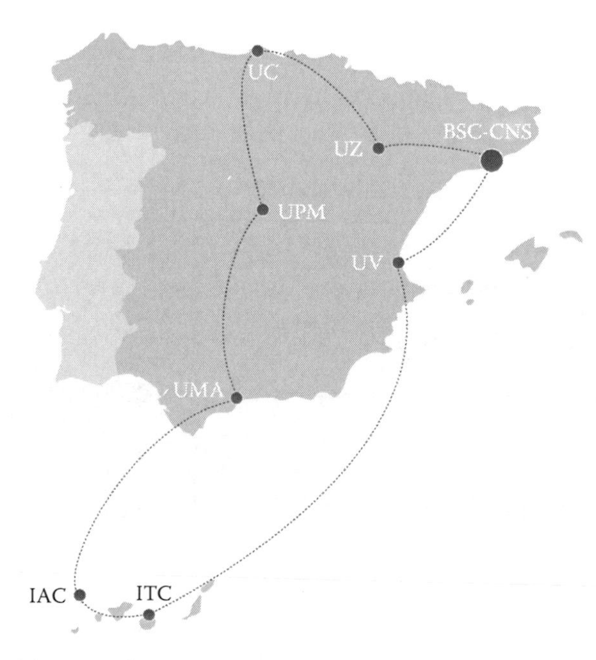

FIGURE 10.2 The Spanish Supercomputing Network (RES) geographic distribution.

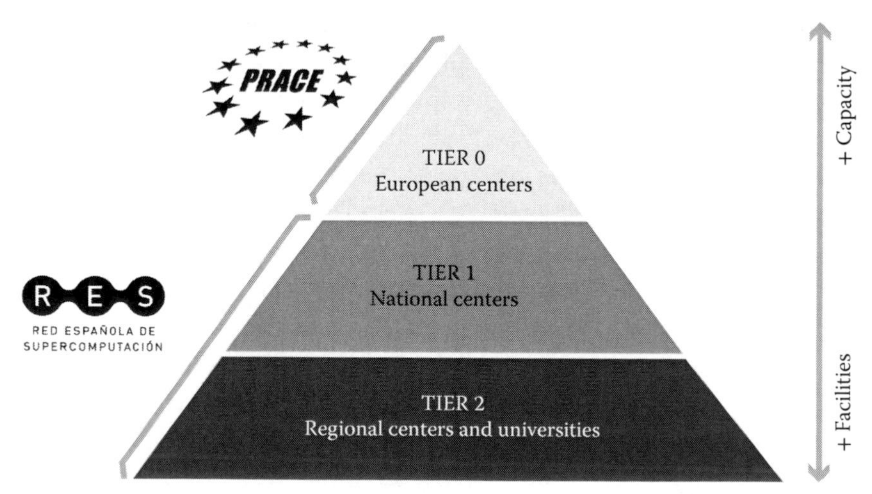

FIGURE 10.3 The Spanish Supercomputing Network (RES) national centers as related to European centers and universities.

Despite the difficult economic situation, long-term planning bore fruit in 2012, with upgrades of several RES supercomputers and a resulting increase in HPC resources available to users. The upgraded supercomputers include MareNostrum (BSC-CNS), LaPalma (IAC), Altamira (UC), Pablo Picasso (UM), Tirant (UV), the new Memento supercomputer at BIFI (UZ), and Atlante (ITC).

All the RES computing capacity offered is made available to the general scientific community via public calls, with applications submitted via a web interface, and are evaluated by a single Access Committee. The allocation of access to the supercomputing facilities is based on criteria of efficacy, efficiency, and transparency.

The evaluation process is mediated by a double filter system, with potential projects first being evaluated by the ANEP (National Agency of Evaluation and Prospective) if they have not been previously evaluated by other relevant national or international institutions, followed by a review by the RES Access Committee. This is composed of a Core Team and four Scientific Expert Panels formed by prestigious scientists external to BSC-CNS and defined according to the classification established by the Spanish Foundation of Science and Technology (FECYT). The budget needed for maintaining the Spanish HPC ecosystem is mainly provided by the Spanish government and local governments from education and research funds.

The expert panels comprise a coordinator, an assistant, and eight reviewers representing the following domains:

- Astronomy, space and earth sciences
- Life and health sciences
- Mathematics, physics and engineering
- Chemistry and materials science and technology

Ever since the benefits of an e-infrastructure in general and HPC in particular, have become obvious, the interest from the Ministry of Economy and Competitiveness has undergone a significant boost. We find other funding sources by the European Union through European joint projects, through contracts with Spanish companies and through technology-transfer funds, as there is an increasing interest in using HPC to expand the market penetration of both Spanish products and services.

On the industrial side, large Spanish companies in the energy, telecommunications, pharmaceutical, manufacturing, banking, and insurance sectors fund their own equipment. There are plenty of R&D research collaborations between large companies and Spanish Universities and Research centers with the participation of BSC and RES R&D teams and facilities.

Therefore, there is a gap to fill regarding HPC use by small- and medium-sized enterprises (SMEs) in Spain, as their computing resources usually amount to just a workstation or a cluster. In order for them to become more successful, we need to offer them both hardware and software services, but also services of modeling, application development, and simulation, among others.

10.4 INDUSTRIAL HPC APPLICATIONS LANDSCAPE

The use of supercomputers is applicable to practically all areas of knowledge. In Spain, it has been applied, without any pretentions to being exhaustive, to the following:

- Physics and engineering: Study of fluid dynamics; design of hulls and appendixes of boats; study of turbulent flows both in airplane wings while flying and inside their turbines; development of virtual wind tunnels; crash, wear and break tests, high-energy physics, and nuclear fusion
- Biology and medicine: Analysis of the human genome; study of the structure and functioning of proteins; modeling of human organs; and development of new drugs and personalized medicine
- Chemistry and materials science: Design of structurally stable nanofibers; design of catalysts; study of reactivity in surfaces; and design of biomaterials, study of combustion processes
- Astronomy and earth sciences: Modeling of climate, hydrological and oceanographic systems; prediction of air quality in the Iberian Peninsula; diffusion of polluting agents; modeling of the emission and transport of natural dust from the Sahara desert to the European continent; study of the impact and the consequences of climate change on a European scale; exploitation of renewable energies; and astrophysics and space exploration
- Economy and humanities: Macroeconomic and microeconomic models; study of human migrations; and archeological modeling

10.5 INDUSTRIAL COLLABORATIONS AND TECHNOLOGY TRANSFER

Apart from its own research and the support for public research, the BSC-CNS joins forces with companies that are leaders of the technological sector to develop innovative solutions, taking into account from the very beginning of their industrial applicability. The BSC-CNS also offers its resources to the business environment as a competitive tool.

One of the main objectives of the BSC-CNS is to proactively transfer technology to industry, both as an objective in itself in terms of dissemination of scientific output, and with the intention to generate industrial returns. Increasing emphasis is being placed by the BSC-CNS management on facilitating interactions with industry at all levels, from direct R&D collaborations to educational activities such as providing technical seminars, and staff exchanges with private industry R&D laboratories.

IT companies and senior executives from companies of different types and sizes visit the BSC-CNS facilities regularly, and are given presentations with examples of usage of HPC in their respective sectors. Some of the industrial sectors covered are aeronautics, automotive, telecommunications, media, pharma, logistics, textile, and governmental IT-related organizations.

The existing collaborative R&D program agreements that the BSC-CNS has with REPSOL, Iberdrola, Microsoft, Intel, IBM, and Nvidia have been extremely active, with significant progress and successful results obtained in multiple projects, as described in the following:

- Collaboration with IBM focuses on three main directions: (1) workload management of data analytics workloads and transactional workloads, (2) high-performance architectures for Big Data with particular attention to the IBM Blue Gene Active Storage (BGAS) architecture and the Parallel In-Memory Database (PIMD) as the key/value store, and (3) power modeling and adaptive data prefetching techniques for the IBM new generation processors.
- The main objective of the Intel-BSC Exascale Laboratory is to conduct research activities on novel programming models and prediction tools that will be needed to exploit extraordinary levels of parallelism in future Intel-architecture-based supercomputers, consisting of millions of cores. Bringing power consumption down to acceptable levels and to make the system fault tolerant are two key

issues that have been investigated, and whose management needs to be transparently handled by highly scalable parallel run-time systems. Future exascale supercomputers will be hugely complex and challenging to understand and to control.

- The BSC-Microsoft Research Centre (BSCMSRC) carries out research activities on transactional memory (TM). BSCMSRC researchers are also exploring the suitability of low-power vector processors for mobile and data intensive workloads. Finally, BSCMSRC researchers are working on the integration of the BSC-CNS StarsS programming model with the message-passing Barrelfish Research operating system from ETH Zurich in Switzerland.
- The BSC-CNS CUDA Center of Excellence (CCoE), in association with the Universitat Politècnica de Catalunya (UPC), recognizes the BSC-CNS's broad-based research success in leveraging the NVIDIA CUDA technology and GPU computing. In addition, research is being conducted in different directions: use of low-power GPUs in platforms oriented to HPC; leveraging CUDA for productive programming in clusters of multi-GPU systems using OmpSs; optimization of applications in different domains—Reverse Time Migration, electronic and atomic protein modeling, and security video surveillance; and the development of software infrastructures to ease the development on GPU-based systems for different system topologies.
- The REPSOL-BSC Research Center is undertaking various research projects to develop advanced technologies applicable to the exploration of hydrocarbons and other areas of interest to REPSOL, such as modeling of subterranean and subsea reserves and fluid flows. The establishment of the center is the result of many years of successful collaboration between BSC-CNS and REPSOL, a highlight of which was the now commercialized Kaleidoscope project, which developed algorithms that enable REPSOL to process subterranean seismic images up to 15 times faster than its competitors, and was voted one of the 5 most innovative projects in the global energy sector.
- BSC-CNS and Iberdrola Renovables are collaborating to design mathematical models to improve the design of Iberdrola's wind farms. The project tackles the extremely challenging simulations

of wind farms. The main objective is to significantly increase the efficiency and power of wind farms by developing numerical techniques to optimize placement of wind turbines.

In addition to the major collaborations detailed above, the BSC-CNS is actively engaged with private industry and government on a range of applied projects. The BSC-CNS has collaborated with more than 100 private companies both nationally and internationally, either in direct R&D collaborations or via European or nationally sponsored project consortia, for example.

- The joint collaboration with Xilinx has undertaken research on OmpSs support for FPGA accelerators. Researchers from the Computer Sciences Department developed support for FPGA accelerators based on the Xilinx leading-edge technology.
- The collaboration with the European Space Agency (ESA) is structured around two projects: Architectural solutions for the timing predictability of next-generation multicore processors, with the objective of enabling the Worst Case Execution Time analysis of time-critical space applications in a multicore execution environment such as the New Generation Multi-Core Processor (NGMP); and Multicore OS Benchmark, with the objective of evaluating the time-predictability behavior of the multicore processors mentioned above and doing research on task-scheduling aspects for multicores (Figure 10.4).

FIGURE 10.4 Logos of companies engaged with BSC-CNS through the years.

The development of operational air quality forecasting and assessment services for various regional governments throughout Spain and for international public bodies are of particular interest, as is the analysis of power generation and other industries' impact on air quality:

- Meteorological Service of Catalonia (SMC): to generate regional climate scenarios at high resolution for Catalonia during the twenty-first century.
- Environment and Water Agency of Andalucía: to provide the Andalucía Government with an operational air quality forecasting and assessment service, which will allow the simulations of photochemical and particulate matter pollution with high spatial and temporal resolution for Andalucía: 1 km^2 and 1 hour?
- International Research Institute for Climate and Society: to enable cooperative efforts between the IRI and the BSC-CNS in areas connecting climate, atmospheric aerosols, and health.
- National Oceanic and Atmospheric Administration (NOAA)—National Centers for Environmental Predictions and Environmental Modeling Center: to develop a new chemical weather prediction system and chemical transport model intended to be a powerful tool for research and to provide experimental efficient global and regional chemical weather forecast.
- AEMET (The State Meteorological Agency): to implement, disseminate, and validate the operational prediction of the North African dust transport in the Iberian Peninsula as well as to perform modeling, detection, follow-up, and characterization studies of atmospheric material.
- Spanish Ministry of the Environment: to develop and implement an operational high-resolution air quality forecasting system for Spain, providing end-users with an air quality forecasting and assessment service for Spain and Europe with higher detail for some hot spot areas.
- Government of the Canary Islands: to develop an information system for air quality forecast and surveillance of the Canary Islands.

10.6 CASE STUDY: REPSOL

REPSOL wants to play an important role in the business of oil extraction, opting for areas with a greater legal certainty, such as the Gulf of Mexico (in the United States) or Brazil. The oil in these areas is difficult to find and extract:

- Depth: 3000 m water + 5000 m land. Approximate cost of extraction: 150 million dollars.
- Subsalt area: salt strata, which do not allow the use of the conventional technology of image generation.

The sediments below the deep and ultradeep waters of the U.S. Gulf of Mexico shelter rich oil reserves, sometimes as much as 3000 meters from the surface. Minerals Management Service (MMS), the federal agency in the U.S. Department of Interior that manages the nation's oil, natural gas, and other mineral resources on the outer continental shelf in federal offshore waters, estimates that the Gulf of Mexico holds 37 billion barrels of "undiscovered, conventionally recoverable" oil. Similar scenarios appear at basins such as Brazil and Angola.

These reserves are very difficult to find and reach due to thick layers of salt that preclude the imaging and visualization of the oil-bearing sands underneath. The oil industry uses sophisticated technologies to locate and visualize these exploratory objectives. These technologies are computing intensive and the success to properly "see underneath" depends largely on the power of the supercomputers used. The increase in computer power makes the application of imaging technologies that until today have been considered as a utopia in the oil industry feasible and allows a more reliable exploration.

The REPSOL-BSC Research Center seeks exploitation of parallel supercomputers for the creation of the next generation of seismic imaging technologies for the visualization of the earth's interior. The REPSOL-BSC Research Center has developed BSIT (Barcelona Seismic Imaging Tools), a software platform designed and developed to fulfill the geophysical exploration needs for HPC applications (more information on www.bsc.es/bsit/index.html). This platform includes different packages for processing seismic data: Forward Modeling, Reverse Time Migration, Full Waveform Inversion, and Electromagnetic Imaging. In addition, the software supports different rheology including acoustic, acoustic with variable density, elastic, and viscoelastic. Moreover, several levels of anisotropy are supported: Vertical Transverse Isotropy/Horizontal Transverse Isotropy (VTI/HTI), Orthorhombic, Tilted Transverse Isotropy (TTI), and arbitrary anisotropy (for elastic and viscoelastic rheology). All this software is available for any kind of computer architecture from general purpose processors to accelerators like GPUs or the Xeon Phi.

The most promising geophysical imaging tool nowadays is Full Waveform Inversion (FWI). Potentially, FWI can retrieve physical parameters for a whole 3D subsurface volume directly from the seismic data. However, the tool has its limitations, and requires specially acquired data (low-frequency, long-offset), good initial models and huge computational power. In BSIT, the FWI tool has inherited the efficient implementation of the algorithmically similar RTM and boosted its capabilities to turn this very costly imaging algorithm into a commodity.

From all this, we can gather how important technology is to gain a good position in this market, necessitating an alliance with new technology generators.

Thus, the BSC-CNS is a strategic ally for REPSOL. Both have created an R&D joint center whose objective is the development of algorithms to be used in the day-to-day production of the company. This joint center has allowed REPSOL to have more advanced technology in image generation, giving it an advantage of 2 years over the rest of the companies.

10.7 STIMULATION PLANS FOR ADOPTION OF HPC BY THE SPANISH INDUSTRY AND SMEs

General collaboration from HPC public centers—specifically the BSC-CNS—has been and still is significant when it comes to large companies (both national and international), but the same cannot be said for projects with SMEs, or for the area of the creation of companies.

The stimulus plan anticipated in the short and medium term is the following:

- Provide SMEs with a competitive advantage via public help for the collaboration projects with the BSC-CNS, funded by these companies
- Generate a sufficient number of start-ups with a capacity for fast growth and a world-wide competitive advantage via a public-private partnership, which extracts economic value to the enormous base of knowledge and technology that is being developed in the BSC-CNS and its environment

10.8 CONCLUSION

Supercomputers can be used in practically all areas of knowledge; from physics and engineering to biology and medicine, as well as chemistry and materials science, earth sciences, and astronomy. It can even be applied to economics and humanities.

We have seen examples of how a supercomputer, through projects of virtual modeling of reality, reduces costs and avoids the risks real experiments would entail, and it even simulates events that cannot be experimentally reproduced.

Are supercomputers a luxury? Clearly not, when we look at all the subjects mentioned above.

Are supercomputers a necessity? In order for supercomputers to become a tool that provides an answer for the needs of our society, two conditions must be met, which are evident but not easy: deploying a critical mass of researchers, developers, and entrepreneurs that increase their competitive advantage thanks to the use of supercomputers, and also having a human staff that is able to help them, with a deep knowledge of the design and the operation of the supercomputers. We sincerely believe that this is the case for HPC in Spain.

CHAPTER 11

Industrial Use of High-Performance Computing in the United Kingdom

Mark Parsons and Francis Wray

CONTENTS

11.1 INTRODUCTION

This chapter discusses the industrial use of high-performance computing (HPC) in the United Kingdom. It is partial and opinionated. For a detailed census of available HPC systems and skills in the United

Kingdom, the reader is directed to http://www.slideshare.net/comth/uk-national-einfrastructure-survey-2014-update-for-hpcsig-may-2014, which is updated annually. Industry in the United Kingdom has a long history of using modeling and simulation and the government has, since the 1990s, encouraged collaboration between industry and academia through a number of programs with varying levels of success and commitment. Despite the reticence with which many U.K. politicians approach Europe, the United Kingdom has also been a key leader in several large European industry HPC programs such as the HPCN TTN (High-Performance Computing and Networking Technology Transfer Node) program and most recently the FORTISSIMO project.

11.2 HISTORY

Since the 1940s, the United Kingdom was in the vanguard of the early use of digital computers to model and simulate the world around us. Much of the early work grew out of the war effort and as such, the use of HPC systems for product development by the defense sector was a key part of early U.K. industrial usage. Such usage continues to this day but these early pioneers have been joined by a broad array of business sectors represented by companies both large and small.

The modern history of HPC in the United Kingdom dates back to the 1980s when the U.K. had an active and innovative hardware industry. It was a period when the U.K. believed it could be a world leader in HPC systems through companies such as ICL (International Computers Limited), INMOS (Inmos Limited, a British semiconductor company), and MEIKO (a pioneer in low-latency scalable systems). Sadly, of course, this was not to be and ICL's Distributed Array Processor (DAP), INMOS's line of Transputer microprocessors and MEIKO's Computing Surfaces (CS1 and CS2) are consigned to history. However, this period in the late 1980s and early 1990s remains an important one.

During these decades, vector machines such as the Cray-1 and Cray-2 dominated the worldwide HPC market. For U.K. academia and industry, such systems vastly exceeded available budgets and so cheaper alternatives with similar performance were developed, utilizing the then new computing paradigms of single instruction, multiple data (SIMD) and multiple instruction, multiple data (MIMD) parallelism. Of course, the reduced price came at a cost in terms of application complexity and portability. It should therefore be no surprise that the United Kingdom was a leading contributor to the message passing interface (MPI) library, MPI 1.0, in the early to mid-1990s. Indeed, at one point in the 1990s, Edinburgh Parallel Computing Center (EPCC) EPCC had written all or part of the MPI 1.0 optimized library implementations for Cray Research, Digital Equipment Corporation, Silicon Graphics, and Sun Microsystems—an early example of successful technology transfer from academia to industry.

At the same time in the early 1990s during Baroness Margaret Thatcher's tenure as Prime Minister, the Government's Department of Trade & Industry (DTI) was persuaded to fund the Parallel Applications Programme. This was a technology transfer program designed to broaden the uptake and effective use of HPC modeling and simulation on parallel computers by encouraging collaboration between industry and academia through the creation of four parallel computing centers. Of these, only two remain today—the Parallel Applications Centre at Southampton University (now named IT Innovation and no longer focusing on HPC) and EPCC at the University of Edinburgh (now one of Europe's largest HPC centers with a strong business collaboration activity for the past 24 years).

The Parallel Applications Programme model has been replicated several times in the United Kingdom and Europe. The University centers received funding to work with companies in the United Kingdom, either introducing them to parallel computing completely or taking their existing software applications and parallelizing them.

At the end of the program the DTI review found that:

- Structuring of the program on four centers had distinct advantages in terms of critical mass and development of staff.
- It left a legacy of functioning parallel application centers. These are continuing, in various forms, with funding from public and private sources.
- The program generated a broad range of company projects, a large number of which were technically successful with significant technology transfer.

The program ran for 4 years, initially entirely funding such projects and then funding a reduced percentage of the total project costs in the latter years. The idea of course being that the centers should become self-sustaining by the end of the fourth year. This initiative was important because it laid the groundwork for later programs.

11.3 THE U.K. HPC ECOSYSTEM TODAY

Coupled to the long-term growth of industrial use of HPC have been periods of commercial supply-side development where a broad range of companies have developed new HPC technologies or services. Although many of these have disappeared, today the United Kingdom has an active and growing HPC supply side—of both HPC products and services. In terms of HPC products, a number of U.K. companies supply HPC cluster systems built from commodity components—a typical example would be Viglen. These companies are complemented by a number of more esoteric system providers who mostly focus on accelerator technologies—particularly field-programmable gate array circuits (FPGAs). Examples of such companies are Maxeler and Alphadata Parallel Systems. The United Kingdom is also lucky to be the headquarters of some major electronics companies—notably ARM Holdings and Imagination Technologies—and storage companies such as Xyratex.

Complementing these hardware companies are a wide spectrum of companies that supply software into, or provide services using such software, to industry and academia. On the HPC tool's front, Allinea develops and sells a widely adopted DDT (Distributed Debugging Tool), whereas a number of specialized compiler companies supply niche markets. In the engineering sector, a large number of specialized consultancy companies provide modeling and simulation services to their clients. A number of Independent Software Vendors (ISVs) are based in the United Kingdom—a well-known example being Numerical Algorithm Group (NAG), which also provides HPC consultancy services to industry.

Coupled with this expanding company base, the United Kingdom has a very active computational science community that acts as both supporters and users of the products and services provided by the supply-side companies. Since the mid-1990s, the government has understood the importance of funding one or more national HPC services and, crucially from an industrial point of view, has always encouraged the host organizations to engage with industry. In the last few years, an ecosystem of HPC systems has grown up, funded from a variety of sources. Many Universities have their own HPC systems, a small number of regional HPC systems have been funded using e-Infrastructure funding and these systems provide a suitable "ramp" onto the large-scale National Services. Although this pyramid structure has been arrived at somewhat accidentally, over the past 5 years it has proven particularly effective in supporting the U.K. academic computational science community and the companies that engage with it.

In contrast to similar publicly funded systems in Europe, U.K. organizations that host these systems are encouraged to engage with industry and may charge for their services. This is provided the rates at which they charge for access are a fair market rate, which does not distort the market, and all relevant taxes (such as VAT) are charged. Although it is often assumed that a center such as EPCC sells considerable amounts of access to its systems, this is not the case. For example, in most years, considerably less than 5% of EPCC's turnover relates to industry buying access on its systems. This is a common picture nationally.

However, most of the time industry collaboration goes beyond the simple sale of access to resources and encompasses both access to HPC systems and the modeling and simulation skills embedded in the universities or laboratories hosting the systems. More often than not, the access to the resources is a small component of a project and is generally related to application software development rather than production runs.

11.3.1 Interacting with Companies in the United Kingdom

The Parallel Applications Programme required the HPC centers to learn how to sell to industry—something many such centers had never done before. Today, EPCC is the largest center remaining from this period—its long-term success is largely due to what it learned in this and subsequent programs (such as Europort and the HPC Technology Transfer Node network funded by the European Commission in the 1990s).

At the same time, regional economic development funding has proven very useful to develop interaction between HPC centers and industry in the United Kingdom. EPCC's Supercomputing Scotland programme, Loughborough University's HPC Midlands activity, and HPC Wales are excellent examples of how to engage with the local business community. In common with many parts of Europe, a large number of local companies are small- to medium-sized enterprises (SMEs). Although a small number of such companies are accomplished users of HPC, the vast majority are not and many do not even think it is relevant for them. Engaging with such companies is a key skill and many of the National and Regional HPC centers in the United Kingdom, including National Laboratories such as Daresbury Laboratory and the Hartree Centre, have learnt how to do this successfully.

11.3.1.1 Developing a Business Mindset

In the United Kingdom, the centers that have developed successful industry engagement programs have all learnt the hard way that to do this you need more than a fast supercomputer and a desire to work with industry (be that for financial gain or to meet the demands of a funding body—or both, of course). There are three critical factors to success:

1. Business development: A properly managed business development process is very important. Companies want to know what they are buying and what they can expect from engaging with the HPC center. Part of this process is psychological. A well-produced brochure and a business suit confer a warm glow of trust. But a well-managed process with a business plan and targets are vitally important if industry engagement is not always going to play second fiddle to academic use of the HPC system.

2. Understanding the problem: If a company is interested in collaborating, it is quite often because they have a complex problem to solve. It is vital, therefore, that a business development manager is supported by technical staff who can properly understand that problem, how long it will take to solve (or if it is even feasible) and can agree on this with the company.

3. Delivery: Scale is important. Companies expect to access expertise within an HPC center quickly. Waiting to hire a new research assistant is not part of their plan. Here, the size of an HPC center is important. For less than 20 people, it is very difficult be flexible. For more than 50 people, it is easy to find staff but it is vital that proper software development and project management processes are used—something it took EPCC, for example, a long time to get right.

Of course, the business community is not homogeneous. Although most interactions are straightforward and positive, all business development staff has stories of terrible meetings where, for example, it has become clear that the managing director of the company is drunk or their Business Directory has directed them to a meeting with a "manufacturing" company that turns out to be a baker!

11.3.2 Two Examples of Industry HPC Projects in the United Kingdom

Companies come in all shapes and sizes. Although larger companies may engage in longer projects, for example 12–18 months as opposed to 3–6 months for a small company, the needs of these companies are surprisingly similar when engaging with HPC skills and services hosted in the academic sector.

11.3.2.1 Integrated Environmental Solutions Case Study

Integrated Environmental Solutions Ltd (IES) is a world-leading provider of software and services focused on making the built environment more energy efficient. By modeling how buildings heat and cool themselves, the company provides its customers with advice that allows them to use energy resources more efficiently and reduces CO_2 emissions.

Architects and designers use IES's SunCast software to analyze the effect of sunlight and shadows on the thermal behavior of buildings. By understanding the effect the sun has on buildings, their thermal performance can be optimized to ensure a comfortable environment for the building's inhabitants whilst minimizing energy use for cooling or heating.

In contrast to the academic world, which is predominately focused on the Linux operating system, many industry applications including SunCast run on Microsoft Windows. In a 4-month project, EPCC modified the code to make use of Microsoft MPI and task-based parallelism. This has reduced runtimes significantly. A particularly complex analysis was reduced from 30 days runtime to 24 hours. This is not complicated parallelism, but the benefits have proven highly valuable to the company, as it now offers SunCast through the cloud as a pay-per-use service.

11.3.2.2 PROSPECT Case Study

PROSPECT started life in 1999 as a small engineering consultancy company in Aberdeen providing design and analysis services to the world's energy industries—particularly in its early years to the oil industry in Aberdeen. It grew rapidly and is now part of Superior Energy Service, a 12,000-employee global company with annual revenues of $6.2 billion.

In common with IES, PROSPECT continually wants to explore new business opportunities and push the envelope of modeling and simulation. EPCC worked with PROSPECT in 2009 as part of the groundwork for the

Supercomputing Scotland initiative to explore a new business opportunity in the renewable energy market. At this time, a large number of engineering projects are exploring seawater wind turbine technology. Modeling these turbines is complicated because they have the wind, waves, and tidal forces acting on them. The turbines are located on towers, anchored to the seafloor and these towers must resist the forces of waves crashing into them, tidal flows, and the gyroscopic effects of the large turbine blades powered by the wind.

In a 6-month project, EPCC worked closely with PROSPECT to demonstrate how to model these devices by coupling a finite element code (ABAQUS) and a computational fluid dynamics code (STAR-CCM+). At the time of this work such a complex simulation had not been undertaken in the renewables sector and this work gave PROSPECT a competitive edge in their marketplace.

11.4 DELIVERING ECONOMIC BENEFIT

For many years, the HPC communities in each country told their governments that by simply buying the next greatest HPC system, there would be direct economic benefits to their country's industry and commerce. In the United States, this has been captured in the slogan, "*To out compute is to out compete.*" As National HPC systems become larger and more complex to meet the increasing needs of science it could be argued that their value to companies is diminishing. The two examples given earlier used very modest numbers of processor cores and this is typical of many industry projects.

In 2007, when the HECToR National HPC service opened at EPCC, it was decided to approach the Scottish government's economic development agency, Scottish Enterprise, for support for a new program to drive increased use of HPC by Scottish companies. However, while reviewing the literature on the industrial use of HPC jointly with Scottish Enterprise, we realized that many claims about the economic benefits of HPC investment are made but these statements are rarely backed up with anything other than examples of successful projects.

Before agreeing to fund the project, which was eventually named *Supercomputing Scotland*, Scottish Enterprise decided to co-fund three pilot projects: the project with PROSPECT described above, a project to model a mud-powered reaming device for the oil industry made by Deep

Casing Tools Ltd, and a genomics service offered by FIOS Genomics Ltd. These projects ran in 2009 and 2010 and were assessed by independent economic consultants once they were completed. On the basis of a detailed, confidential analysis of the companies' economic outlooks, the analysis concluded that, "the return over the 3 years to 2013 would appear to be good generating approx. £10 net Gross Value Added* for every pound of public spend with a five year projection of £25. Clear benefits have been derived by the firms in the pilot." This was an excellent result but it should be borne in mind that it was a small sample size (three projects) and was based on relatively small investments by Scottish Enterprise (on average £35,000). Such small investments and the fact that the projects were chosen as demonstrators, can lead to high impact ratios. Nevertheless, this is one of the few occasions a detailed attempt has been made to measure the benefit that adopting HPC and working with the skilled staff in an HPC center can bring to a company's bottom line.

11.5 FACTORIES OF THE FUTURE: FORTISSIMO

In the European context, the PlanetHPC project helped to make the case for European Commission intervention similar to the Supercomputing Scotland initiative and building on industrial HPC activities in a number of European Union countries. The Fortissimo project, part of the *Factories of the Future* program, brings together Europe's leading HPC providers who work with industry and a set of Experiments with companies, predominately small- to medium-sized enterprises (SMEs), to demonstrate the benefits HPC can bring to their company. These experiments are larger than Supercomputing Scotland projects, up to €250,000 in funding, and generally bring together three or four organizations to tackle a business challenge.

The project started in July 2013 with 20 experiments and has recently added a further 20. By the end of the project in 2016, the project will have delivered approximately 60 experiments and will have around 90–100 partners. Led by EPCC and with total costs of €22 million and funding of €16 million, this project is a major investment by the European Commission in HPC for industry.

* Gross Value Added is a widely used economic indicator and is calculated by subtracting the cost of inputs from the company's turnover.

11.6 CONCLUSIONS

This chapter has sought to give an impression of how industry and academia work together in the United Kingdom to each other's benefit. A number of lessons have been learnt over the past 25 years of engagement between industry and HPC centers in the United Kingdom:

- HPC can bring real benefits to U.K. industry and these accrue just as easily to small companies as they do to large companies.
- HPC centers must deliver a high-quality product with highly skilled staff, clear project goals, and a professional outlook.
- Cycle sales only make up a small part of what U.K. industry wants to buy. Industry's main need is skills—and this is true if they are purchasing from the academic or commercial sectors.
- Even today, many companies are afraid to make their first step into the world of HPC. The hurdles are still high and programs such as Supercomputing Scotland and Fortissimo are vital in helping to lower those barriers.

However, when initial projects are successful, the benefits to the companies who then adopt HPC as part of their day-to-day business processes are large and sustained.

CHAPTER 12

U.S. Industrial Supercomputing at NCSA

Merle Giles, Seid Koric, and Evan Burness

CONTENTS

12.1 INTRODUCTION

The beginnings of industrial supercomputing at the National Center for Supercomputing Applications (NCSA) are rooted in a rising sentiment during the 1980s that America had to maximize the benefits found at the intersection of science and industry to compete. National security concerns borne out of World War II and the Cold War resulted in funding for supercomputers at the Departments of Defense and Energy, leaving open-science and open-engineering experts limited access to advanced computational resources and expertise. These concerns are described more fully in Chapter 1.

NCSA's funding began in 1986 as an outcome of the 1982 Black Proposal submitted by Dr. Larry Smarr and his colleagues (see Chapter 1). Included in the initial strategy was the establishment of the Industrial Partners Program, later named the Private Sector Program (PSP). Since 1986, the PSP has partnered with and provided consulting advice to nearly one-half of the U.S. FORTUNE50,® in a range of sectors including manufacturing, oil and gas, retail/wholesale/distribution, health care, pharmaceuticals, agriculture, food, information technology, bioinformatics, transportation, finance, and more. PSP's participation in NDEMC (see Chapter 21) led to an additional and continuing investment in experts trained in the use of engineering applications in supercomputing environments.

A selection of large-company partners, both current and former, include

- Agriculture: ADM, FMC Corporation, Syngenta
- Finance: JPMorgan, Morgan Stanley
- Information and Communication Technology, including software: AT&T, Cray, Dell, IBM, Intel, Microsoft, Nokia-Siemens Networks, American Airlines Sabre Systems, Schlumberger, Simulia Abaqus, Sony
- Insurance: Allstate, State Farm
- Manufacturers: Boeing, Caterpillar, Deere, Dow, Eastman Kodak, Ford, GE, Honeywell, Motorola, Procter & Gamble, Rolls-Royce, United Technologies
- Medical and Pharmaceuticals: Abbott Labs, Eli Lilly, Mayo Clinic
- Oil and Gas: BHP Billiton, BP, ExxonMobil Upstream Research, Phillips Petroleum
- Retail, Wholesale, and Distribution: Kellogg, Sears

An expectation routinely heard from others is that large companies have enough money and power to solve all technical challenges themselves. After all, they have fleets of PhDs and their own supercomputers, so why should governments and national agencies bother with collaborating? The answer is simple: large companies have enough expertise to know what the problems are and what is not getting done with existing science and engineering. In other words, it is these companies that push the very boundaries of science and engineering in ways that no one else can.

To qualify that statement a bit, perhaps we should note that large companies are very focused on product commercialization, and they will invest in R&D in ways that have rather immediate and obvious returns on investment. Or, stated a bit more directly, they will invest in ways that improve profitability. Less obvious and immediate returns are scrutinized more closely because the potential ROI upside carries with it additional risk.

Collaborating with supercomputer centers, whether at universities or federally-funded laboratories, helps companies lower risk. Spreading risk among collaborators, or buying down risk as fast-followers of innovation, is good business. This holds true in the scientific and engineering aspects of production and commercialization, and also in the technical solutions to science and engineering. Thus, while the world's largest corporations take on some risk in their own centralized R&D labs, they also pay close attention to advances in supercomputing efforts happening elsewhere (see Chapter 13).

12.2 SCIENTIFIC NETWORKS AND EARLY CHALLENGES WITH DATA

Digital simulation has always produced lots of data to be analyzed and visualized. Bob Wilhelmson, a co-Principal Investigator with Larry Smarr on the Black Proposal, studied severe storms that produce the strongest tornadoes. Like many campus researchers in the early 1980s, he needed to travel to the National Center for Atmospheric Research (NCAR) in Colorado to use its large computing systems. He would submit jobs to NCAR either via telephone modems or punch cards and wait for output in the mail.

At the time, Wilhelmson's outputs were two-dimensional contour plots of his NCAR data that were written to microfilm and mailed. As volumes of data increased, Wilhelmson influenced decisions that resulted in NCSA doing things a little differently; specifically, newer and faster. This included animating the data using software developed around animation techniques from filmmakers on the West Coast. This led NCSA's Donna Cox and her team to produce scientific visualizations of severe storms that would ultimately be shown in planetariums, museums, and more recently, in film and television documentaries such as *Cosmic Voyage*, *Hubble 3D* in 2010, and *The Tree of Life* starring Brad Pitt and Sean Penn in 2011.

Digital visualization of data produced during a simulation impacts research. Teams in the 1980s dealt with millions of pieces of data, and in the 1990s with billions and trillions of pieces of data. Wilhelmson's storm simulations analyzed both storm structure and storm evolution. What was exciting about NCSA's new approach, Wilhelmson claims, was that he could see renderings of the data no matter what its size, and ask questions that could be answered quickly with new animations.

Today, researchers carry out simulations and produce visual animations directly from the simulation data, which is made available through web browsers. Not coincidentally, the roots of modern web browsers came from within NCSA. Initially, NCSA's response to the Ford Motor Co. challenge to develop scientific networks (see Chapter 1) resulted in NCSA Telnet in 1986, followed by NCSA Mosaic™ in 1993. In his book *The World Is Flat*, Thomas L. Friedman credits NCSA Mosaic™ as the "first really effective, easy-to-use Web browser" for Microsoft Windows. NCSA co-developers Marc Andreessen and Eric Bina, without direct funding, built the browser software to enable scientists to use supercomputers that were in remote locations, and connect them via NSFnet. Graphical visualization thus made the use of supercomputing easier to analyze and understand.

In 1994, Andreessen and a number of original NCSA Mosaic™ authors partnered with Jim Clark, the founder of Silicon Graphics, to spin out Mosaic as a separate company, which would quickly be renamed Netscape Communications. Jim Barksdale, former Netscape CEO recalled in Friedman's book that "Every summer, *FORTUNE* magazine had an article about the twenty-five coolest companies around," and "that year [1994] Mosaic was one of them." Ultimately, NCSA Mosaic™ became the foundation of Netscape Navigator and its code descendent Mozilla Firefox, as well as Microsoft's Internet Explorer® through University of Illinois licensing agreements with Spyglass.[1] Friedman credits NCSA prominently as one of the 10 forces that flattened the world by spawning a "new age of connectivity" in the mid-1990s.[2–4]

12.3 NSF AWARD FUNDS STUDY OF SIMULATION-BASED ENGINEERING AND SCIENCE (SBE&S)

NCSA's PSP was awarded $200,000 in 2009 to study and report on industrial SBE&S, the first NSF award of any kind to the PSP. With a core group of friendly manufacturing partners, PSP and IDC (International Data Corporation) conducted a series of surveys, including face-to-face conversations with corporate users of advanced 3D simulation applications. Detailed reports are available on the NCSA industry website www.industry.ncsa.illinois.edu, with select comments and findings shared:

1. Fewer than one in six organizations claimed their applications, as currently constituted, would meet requirements for the next 5 years.
2. More than 3 in 10 responded that the underlying science needs to be improved, yet more than one-half (53%) of respondents admit it is not always easy to distinguish between needed improvements in the underlying domain science and in the related computational and computer science.
3. A total of 30% of organizations named specific ways they have to "dumb down" their problems to complete the runs in reasonable amounts of time, with the most frequently occurring strategies being the use of undesirably coarser meshes and not fully exploiting the known science. This, despite a surprisingly large number of respondents also claiming that increasing the realism in their simulations is nevertheless limited by the underlying science.
4. The main technical limitations preventing the sites from dramatically increasing the problem size or other dimensionality of their simulations were models/algorithms not scaling enough, inadequate latency/bandwidth, inadequate processing power, and, as aforementioned, the need to advance the underlying science.
5. Of the two dominant engineering methodologies used among manufacturers (Computational Fluid Dynamics and Finite Element Analysis), CFD is known to scale rather well using HPC, yet it was observed that CFD simulations at one company scaled to 1000 cores while CFD simulations at another struggled to use 32. It was found that the underlying science differs in that the scalable domain is compressible fluids while the domain that struggles to scale is incompressible fluids. What is not known, however, is whether this is a fundamental barrier in the algorithm or inadequate understanding of the science.
6. New physics regimes are increasingly complex; with the technology of just 5 years ago the computational modeling for these complex projects would have taken 400 years to do the modeling.
7. Realism in biology is unresolved—we need better science.
8. Too many simulations are running on laptops. Attempts to run the same code on HPC results in the code thrashing back and forth

between disk and memory and it runs for weeks. With sufficient memory it might run in minutes.

9. The science in some structural and fluid physics is adequate, but to achieve realism 10 times more compute power is needed in the same time frame. Output is lacking detail, so digital simulations must still be validated with physical testing.

Statements 2 and 9 above conflict and generate confusion. What we know is that capabilities vary according to specific science domains and specific companies, perhaps suggesting that more work needs to be done in the commercial digital engineering community, particularly with commercially available software applications, to increase realism and predictability.

12.4 ENGINEERING-CENTRIC INDUSTRIAL SUPERCOMPUTING

NCSA's PSP decided in 2010 that it could no longer serve its partner communities adequately with supercomputers built for academic science. Although this was seemingly contrary to the original premise of the 1982 Black Proposal, it was actually a response to the maturity of digital engineering design over the past 20 years. In other words, specific attention needed to be given to the unique demands of industry; namely, near-perfect up-time, high-memory nodes, lightning-fast I/O, quick access to resources, and a consulting team that could push performance boundaries and share these learnings with company partners.

iForge (pronounced like iPhone and shorthand for "Industrial Forge") was designed to use enterprise-class hardware specifically designed to serve the I/O and memory challenges of the most widely used commercial structures (FEA) and fluids (CFD) applications. The platform design choice was also predicated on the fact that the vast majority of commercial off-the-shelf (COTS) codes perform on fractions of entire supercomputers, measured in hundreds to thousands of cores, not thousands to hundreds of thousands.

Secondary to the choice of platform architecture was the desire to serve the "power users" in companies in ways that differ from what might be possible on their home company supercomputers. In other words, iForge is supported within a supercomputer center with more willingness to take risk with newer technologies than would otherwise be considered prudent in the vast majority of companies.

By all accounts this has worked splendidly, with "power users" capably climbing the supercomputer performance curve with lots of help and attention. This collaboration has resulted in these power users adopting workflows with more complexity and performance than previously achieved, while having confidence that persistent access to powerful platforms and expertise is available. The platform and expertise combination has achieved a number of application performance breakthroughs in the past year, which are described below.

NCSA's primary takeaway has been that varieties of user demands require varieties of computing design, prompting a pressing need for validated demonstrations of performance with multiple hardware designs and software configurations. The PSP service model has matured to the point that demonstrations are now regularly performed on heterogeneous configurations on iForge and the NSF-funded Blue Waters. Importantly, the PSP team regularly tests performance using differing message-passing protocols, compilers, libraries, and solvers to find optimum winning combinations.

12.5 BLUE WATERS: AN UNMATCHED SUPERCOMPUTER FOR ACHIEVING SUSTAINED PETAFLOPS

The Key Performance Indicator for the world's most powerful supercomputers remains peak FLOPS, or floating-point-operations-per-second. Twice annually the TOP500® list publishes a list of supercomputers ranked by peak mathematical performance using every node available. The application used to measure performance has been a useful reference point for the supercomputing field for several decades.

Today, leaders in supercomputing are taking a more holistic, or balanced, view of performance—one that focuses on processing power, but also on the ability to feed, receive, and analyze the all-important data users bring to and generate on these systems. How well a system performs on all of these important tasks is perhaps most appropriately measured as a function of sustained performance on real-world applications.

The Blue Waters Project at NCSA aims to embody this perspective in every possible way. Though it has indeed proven to be among the fastest systems in the world as determined by theoretical peak FLOPS, achieving top status on this benchmark has not been an aim of NCSA nor the NSF. Instead, Blue Waters was conceived as and remains focused on being the first system to deliver sustained performance of one petaflop or greater for real-world applications. In other words, the story of Blue Waters is not one of peak speed, but of end-to-end application productivity.

With that dual mission of large-scale science and a broad range of disciplines, Blue Waters was designed to be a well-balanced system, capable of fast computing, with a fast interconnect and immense data capabilities. Summary specifications include

Total compute cabinets (racks)	288
Sustained performance	>1 petaflop
Total peak performance	13.34 petaflops
Total system memory	1.476 petabytes (1,476 terabytes)
Total online usable storage	26 petabytes on disk (26,000 terabytes)
Usable nearline tape storage of data	300+ petabytes (300,000+ terabytes)
Aggregate I/O bandwidth	>1 terabyte/second

With Blue Waters, University of Illinois researcher Klaus Schulten and his collaborators determined, for the first time, the chemical structure of the entire protein capsid that encases HIV (the virus that causes AIDS). The unfolding of this capsid is a possible target for intervention in the infection process. Schulten's team simulated 100 million atoms to accomplish this, yet the requirement to study living cells rather than dead ones is expected to require one billion atoms. Beneficiaries of billion-atom computational science include antibiotics, viral infection processes, and next-generation biofuels. Supercomputers, indeed, are the final instrument for solving cell-scale structures (Figure 12.1).

FIGURE 12.1 **(See color insert.)** May 30, 2013 cover of *Nature* magazine.

12.6 ENGINEERING SIMULATION PERFORMANCE BREAKTHROUGHS

Structural Mechanics: The combination of a purpose-built supercomputer for moderate- to large-scale engineering (iForge) and an extreme-scale, well-balanced supercomputer (Blue Waters) has offered tremendous opportunities to test the capabilities of a wide variety of applications. The PSP domain team has recently focused its attention on commercial off-the-shelf (COTS) used by corporate engineering teams, primarily in FEA and CFD. COTS applications were chosen for at least these specific reasons: (1) ready access through academic and commercial licenses, (2) wide-spread usage by manufacturers large and small, and (3) COTS developers rarely have access to supercomputers of any scale, let alone those of extreme-scale, which therefore limits their development opportunities.

Livermore Software Technology Company's LS-DYNA, an explicit FEA code used for simulations in the auto, aerospace, construction, manufacturing, military and bioengineering industries, recently scaled to 8,000 and then to 15,000 cores on BlueWaters—a world record at the time for scaling any commercial FEA code.

Dr. Seid Koric, a Technical Program Manager with NCSA's PSP and a University of Illinois adjunct professor of mechanical science and engineering, led the collaboration team by working with the hardware vendor, the software developer, and a large manufacturer. Using a model with 26.5 million elements, or particles, LS-DYNA quickly scaled to 1000 cores on iForge (left line), and then to 2000 cores on Blue Waters (center line) (see Figure 12.2). Koric continued to run larger and larger real-world problems on Blue Waters, pushing the code ultimately to 8,000 cores.

Progress was iterative, with repeated analysis of performance bottlenecks addressed by the entire collaboration team. Koric worked with both the hardware and software vendors to efficiently distribute the problem across memory and look for ways to improve communication overhead. The collaboration worked, with key participation by the vendors, the manufacturer, and PSP. Ultimately, runtime decreased nearly 50% from almost 12 hours to 6.

Intrigued by these performance gains, a second manufacturer's model with 72 million elements was used to scale to a world-record 15,000 cores in January 2014. The model itself was three times larger than the first, contributing to a more efficient use of memory and cores. Runtime dropped impressively from 20 hours to two (Figure 12.3).

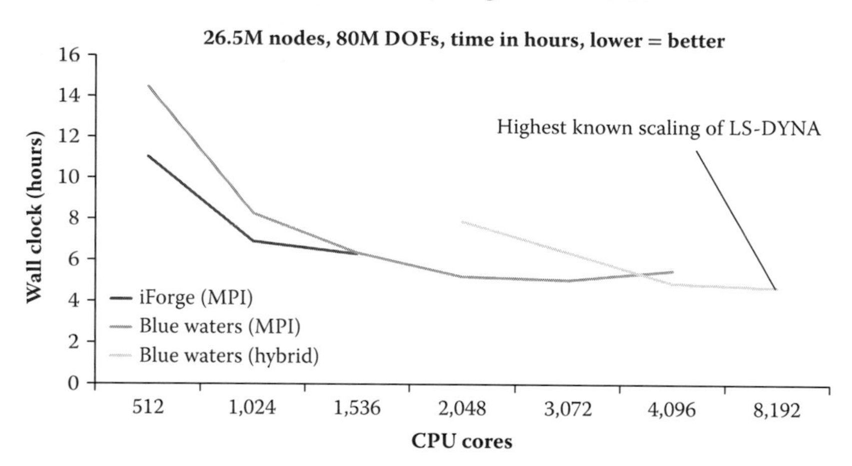

FIGURE 12.2 Graph depicting LS-DYNA scale to 8,000 cores, using a model with 26.5 million elements with 80 million degrees of freedom.

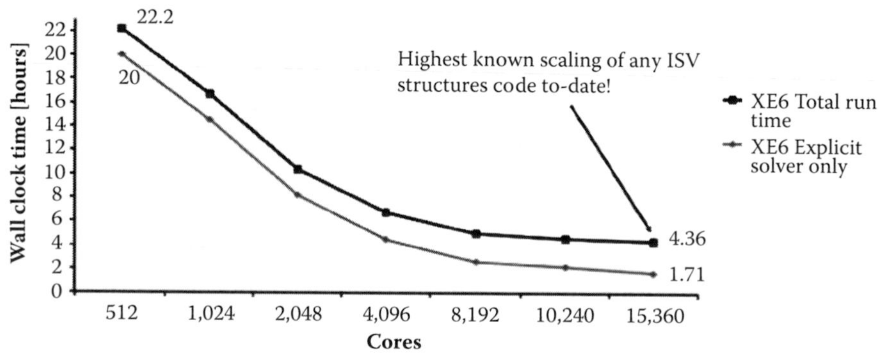

FIGURE 12.3 Graph depicting LS-DYNA scale to 15,000 cores using a model with 72 million elements.

Computational Fluid Dynamics: NCSA's success story with LS-DYNA garnered the attention of other industry partners and software vendors, each looking to push the performance envelope of digital modeling and simulation. One such software company was ANSYS, which offered sufficient licenses to attempt to run Fluent, its CFD code, on tens of thousands of cores.

Fluid dynamics is a branch of fluid mechanics that uses numerical methods and algorithms to solve and analyze problems that involve fluid flows. CFD is an incredibly important aspect of exploring physics within the manufacturing community as air and liquid interactions with surfaces must be fully understood. PSP's Dr. Ahmed Taha led the collaboration team, scaling Fluent to 16,000 cores in February 2014 using a 111-million-element model representing airflows around the body of a truck vehicle. When it was determined that further scaling would likely require a more complex problem, ANSYS brought a representative real-world case comprising an astounding 850 million elements; nearly eight times the size of the earlier case. The new model represented a specific type of combustor used inside an aircraft jet engine.

The PSP team tapped into some uniquely capable resources to begin addressing this challenge. The 3D model had never been completely assembled, so a special high-memory node was required to stitch together six files into one complex model. The mesh ultimately required more than 570 gigabytes of memory to be built to completion—an amount more than twice what could be accessed in typical nodes elsewhere at NCSA. Having reached this critical model-building milestone, the collaboration team turned its attention to conducting full-physics simulations on Blue Waters. Figure 12.4 represents the performance curve with decreasing time-to-solution all the way to 20,480 cores. Figure 12.5 represents the efficiency at which the code performed, with the uppermost straight line representing perfect scalability. When this was demonstrated in May 2014, it represented the world's fastest, best-scaling commercial engineering

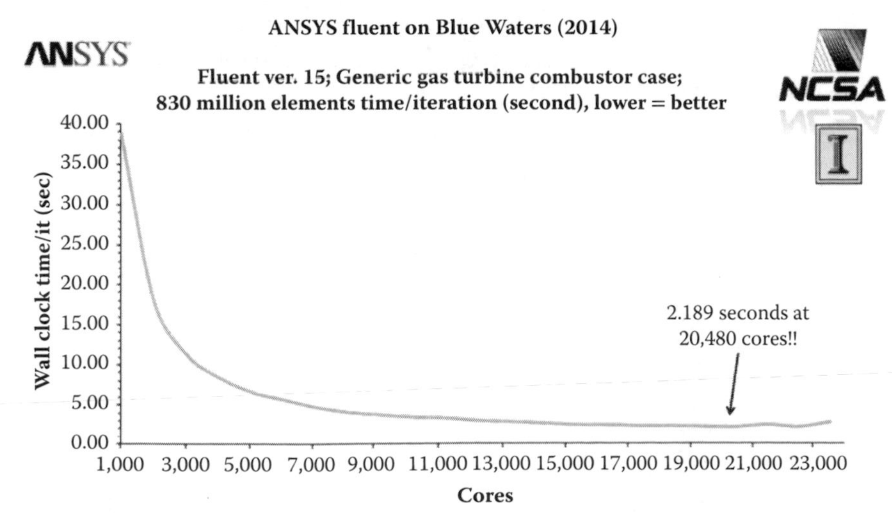

FIGURE 12.4 Graph depicting ANSYS Fluent improving to 20,000 cores.

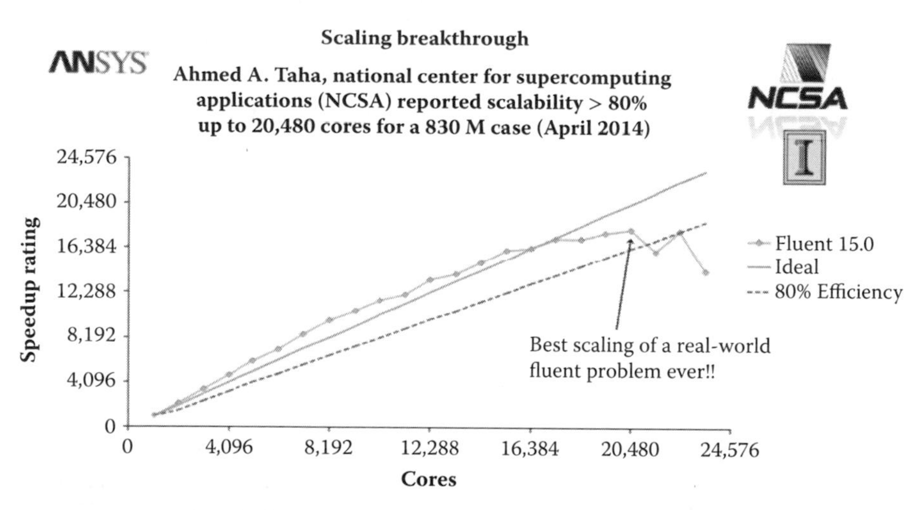

FIGURE 12.5 Graph depicting ANSYS Fluent's near-linear scaling efficiency.

performance of all time. This simulation ultimately scaled to 36,000 cores in November 2014.

Multiphysics: One of the benefits of the International Industrial Supercomputing Workshop (IISW), as initially described in the Executive Summary, is learning where and when to collaborate. A very recent example of such center-to-center collaboration involves Barcelona Supercomputer Center (BSC) and NCSA. BSC's Dr. Mariano Vázquez and Dr. Guillaume Houzeaux developed Alya beginning in 2004, a multiphysics code used to simulate complex engineering problems. Vázquez and NCSA's Giles met in Stuttgart, Germany, at IISW's second annual meeting in 2011, prompting a casual collaboration over several years between BSC and NCSA's PSP.

Although BSC has its own supercomputer on which to develop the code, NCSA has a bigger one, so the teams decided in 2014 to collaborate to push Alya's performance capabilities. BSC provided large multiphysics inputs from incompressible fluid flow in a human respiratory system, a low Mah combustion problem in a kiln furnace, and a coupled electromechanical problem in a human heart (Figure 12.6). It took more than 2 months to port and optimize Alya to run on the Cray-specific Blue Waters, but the work paid off by achieving unprecedented levels of scalability for a multiphysics code.

Figure 12.5 shows the drop-off in scalability for Fluent beginning at approximately 18,000 cores, ultimately achieving 80% efficiency before no more performance gains could be achieved. Alya achieved nearly

100% parallel efficiency running the human heart simulation (Figure 12.7), and nearly 90% efficiency for the kiln furnace model (Figure 12.8), contradicting a commonly held belief that engineering simulations do not scale efficiently on large supercomputers, thus opening a new horizon of scalable applications of importance to industry.

It was expected that Cray's proprietary interconnect played a key role in achieving this performance. Initial testing was promising, scaling to 25,000 cores. Continued testing in April 2014 ramped the code to an astounding 100,000 cores (Figure 12.9). Such scaling and efficiency

FIGURE 12.6 **(See color insert.)** Static numerical model of a human heart.

Alya – human heart

BSC "Alya" on NCSA blue waters; 3.4 billion elements
non-linear solid mechanics coupled with electrical propagation

Speedup
120,000
100,000
80,000
60,000
40,000
20,000
0
CPU cores
Alya achieves near 100% scaling efficiency at 100,000 cores!!
100,000
65,536
32,768
Ideal
427M wrt 1k
427M wrt 32k
3.4B wrt 32k

FIGURE 12.7 3.4 billion elements were in the human heart model, which coupled solid mechanics with electrical propagation. Two smaller models scaled to 100,000 cores, but with less efficiency beginning at 32,768 and 65,536 cores.

Alya – kiln furnace

BSC "Alya" on NCSA blue waters; 4.22 billion elements transient incompressible turbulent flow coupled with energy and combustion

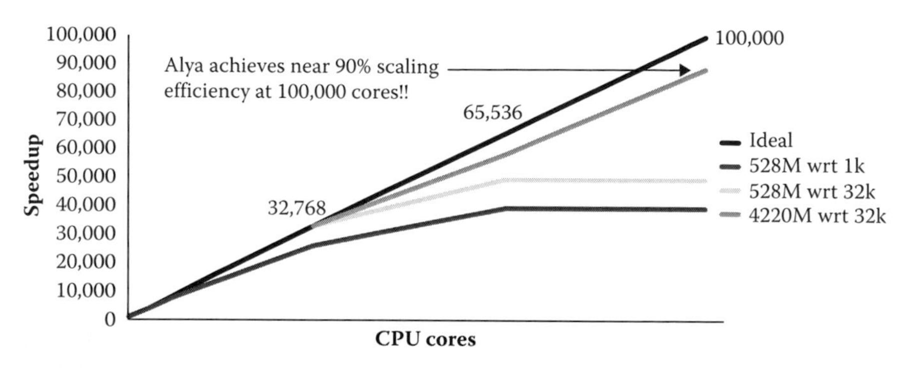

FIGURE 12.8 4.22 billion elements were in the kiln furnace model, which coupled transient incompressible turbulent flows with energy and combustion. Two smaller models scaled to 100,000 cores, but with less efficiency beginning at 32,768 and 65,536 cores.

3D view of extreme scale simulation

10 largest jobs running on Gemini torus – March 2014

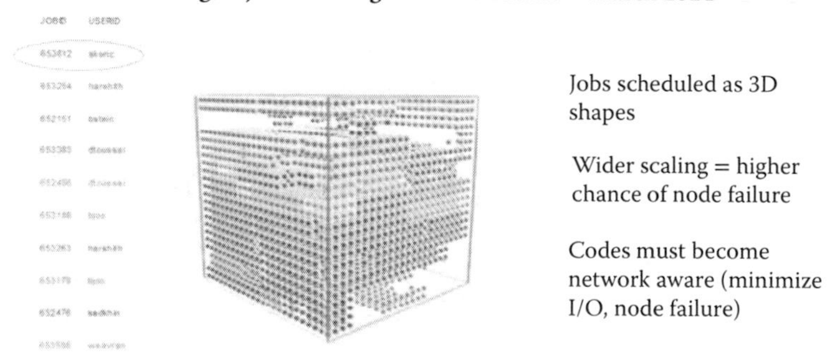

FIGURE 12.9 **(See color insert.)** Graphical depiction of node utilization on Blue Waters, and how network and topological awareness is required to optimize applications.

is raising significant interest in evaluating Alya further, as well as comparing other available multiphysics applications. The promise of more quickly completing complex simulations will impact the management and understanding of human health and other physical phenomena as well as the design of complex machines.

12.7 DIGITAL MANUFACTURING AND DESIGN INNOVATION INSTITUTE

On February 25, 2014 U.S. President Obama announced a $70 million award for a Digital Manufacturing and Design Innovation Institute (DMDII) as part of a White House initiative known as, the National Network for Manufacturing Innovation. Matching funds from corporations and universities brought the institute's total investment over 5 years to $320 million. The awardee was the Chicago-based non-profit organization UI LABS, a technical research center linked to the University of Illinois. NCSA PSP's leadership and its manufacturing partners provided key foundational vision for the DMDII award, citing the need for advanced large-scale digital capabilities to adequately integrate manufacturing and virtual simulation data. One of DMDII's tasks is to bridge the manufacturing sector's existing digital/physical divides and develop a unifying "digital thread" that weaves together all aspects of the manufacturing supply chain.

12.8 SUMMARY

Digital scale and complexity reaches deeply into organizations and entire industrial sectors. Supercomputers are ultimately the only resources capable of approaching realism in the manufacturing and biomedical sectors, to name just two, due to the limitations and costs of physical laboratories (see Chapter 19). NCSA has been deeply involved in providing value to highly sophisticated user communities, and has been privileged to foster changes in workflow and real returns on investment for its partners and clients. The center has been able to build valuable intersections between and among companies in single sectors, and has also leveraged similarities between sectors in ways that companies themselves would not naturally accomplish. Indeed, the core competencies of supercomputer centers around the globe are central to accelerating discovery, invention, and economic innovation.

REFERENCES

1. Clark, Jim (1999). *Netscape Time: The Making of the Billion-Dollar Start-Up That Took on Microsoft*. St. Martin's Press, New York, NY, ISBN 978-0-312-19934-0.
2. Friedman, Thomas L. (2007). *The World is Flat: A Brief History of the Twenty-first Century*. Picador/Farrar, Straus and Giroux, New York, NY, ISBN 978-0312425074
3. NCSA press releases, http://www.ncsa.illinois.edu/news

4. Vázquez, M., Houzeaux, G., Koric, S., Artigues, A., Aguado-Sierra, J., Arís, R., Mira, D., et al. *Alya: Towards Exascale for Engineering Simulation Codes*, Technical paper submitted to International Supercomputing Conference 2014, April 22, 2014, Cornell University Library, http://arxiv.org/abs/1404.4881.

FIGURE 6.3 Luna Rossa AC72 “flies” over the water during a training session.

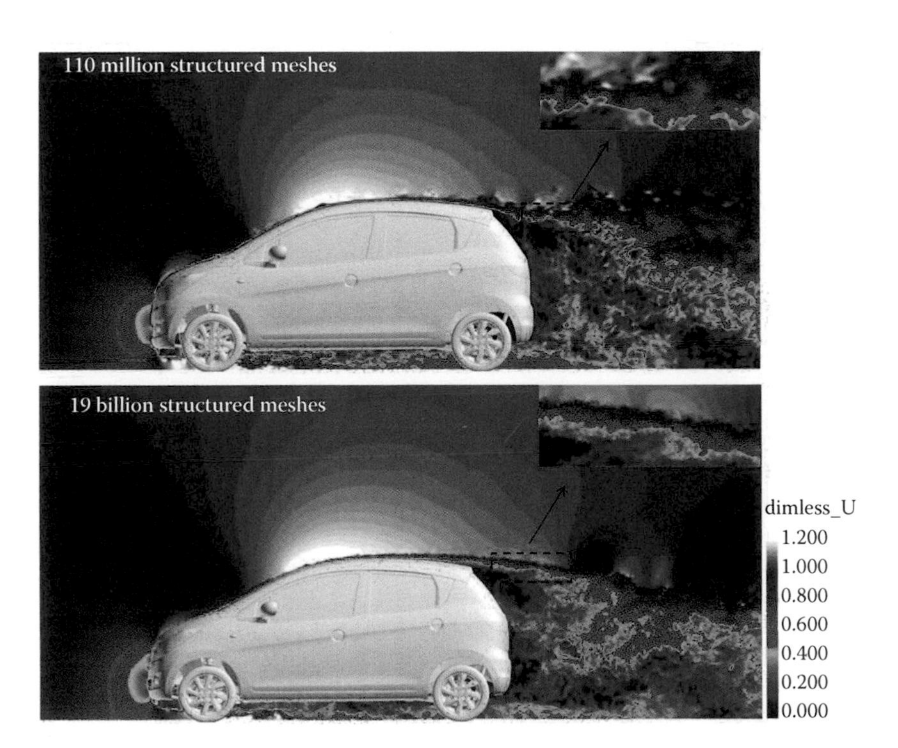

FIGURE 7.5 Next-generation vehicle aerodynamics simulation (RIKEN/AICS and Suzuki Motor Corp.) using 19 billion elements in hierarchical structured meshes.

FIGURE 8.2 3D molecule visualization in the CAVE at SURFsara.

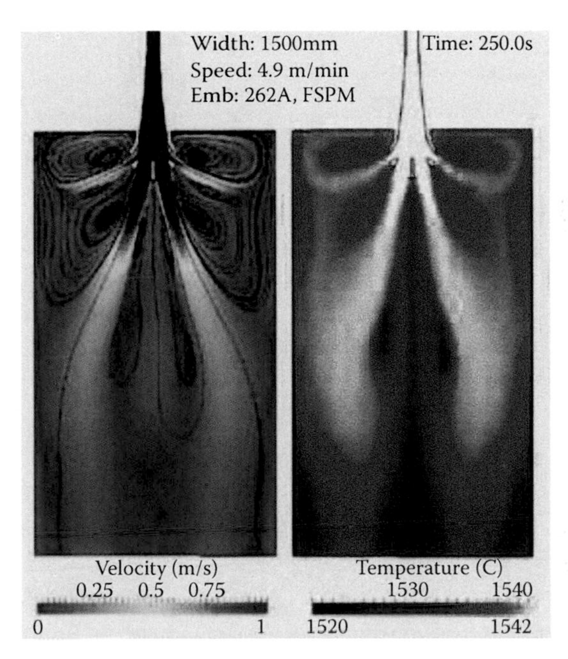

FIGURE 8.6 Temperature and flow in the mold at the Direct Sheet Plant (DSP). This specialized thin casting machine conducts the solidified steel from the mold directly to the rollers, and can reach very high casting speeds. Good quality numerical simulations are required to understand the flows.

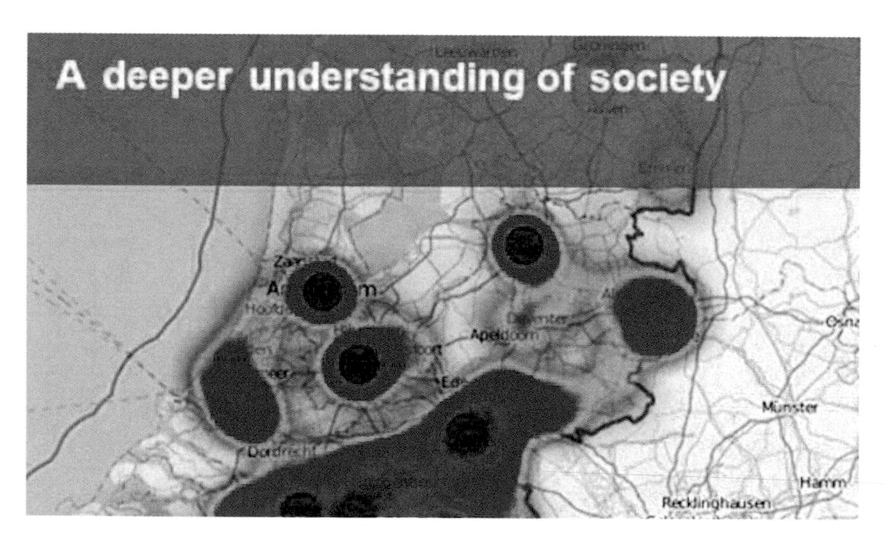

FIGURE 8.7 Occurrence of word "carnaval" in tweets. Carnaval is an event celebrated primarily in the southern provinces of The Netherlands. The results show where Carnaval is a hot topic on tweets, and enables retail stores to adapt their products in this period, for police to concentrate manpower in the right places, etc.

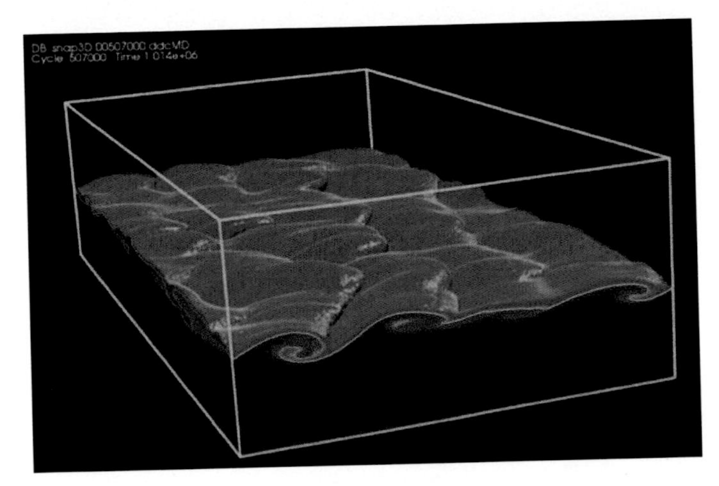

FIGURE 14.1 TriLabs scientists predicted wave action caused by sheer between two fluids using molecular dynamics simulations on 62 billion atoms (Kelvin–Helmholtz instability). All of the processors on LLNL's BlueGene/L supercomputer ran for 6 weeks to produce this 1-cubic-micrometer simulation of liquid copper flowing like waves reaching a shore across liquid aluminum.

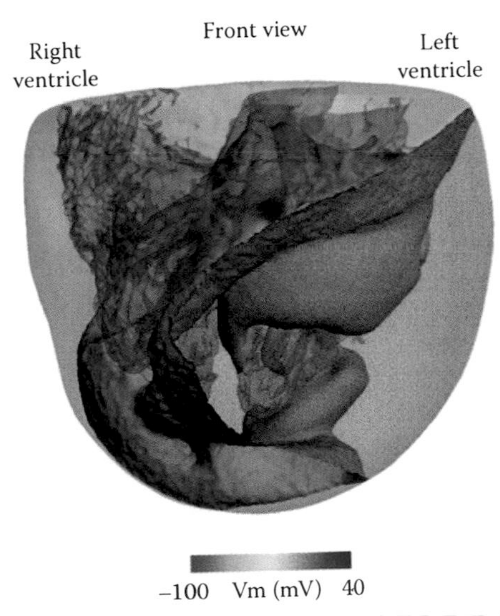

FIGURE 14.2 The Cardioid code developed by a team of Livermore and IBM scientists divides the heart into a large number of manageable pieces, or subdomains. On Livermore's Sequoia supercomputer, the code ran near cellular-level resolution at 60 heartbeats per minute.

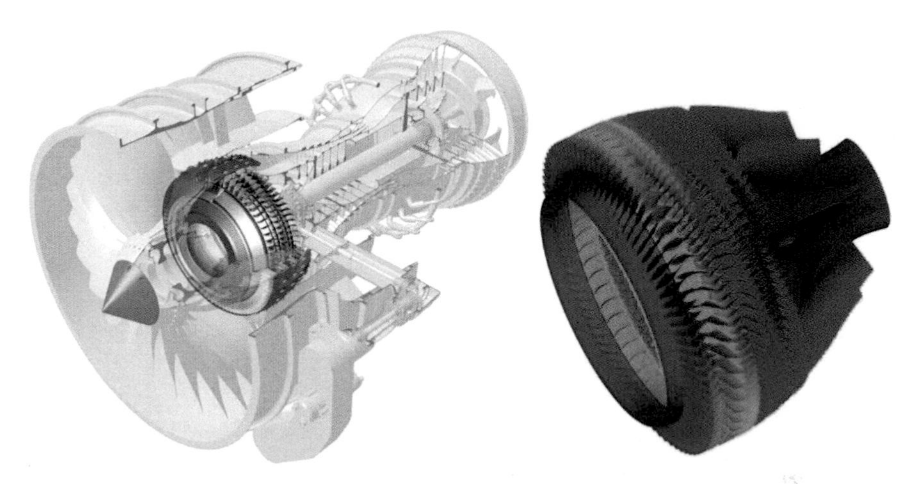

FIGURE 8.8 Low-pressure compressor module in a jet engine and example of simulation result (courtesy of Techspace Aero [safran].)

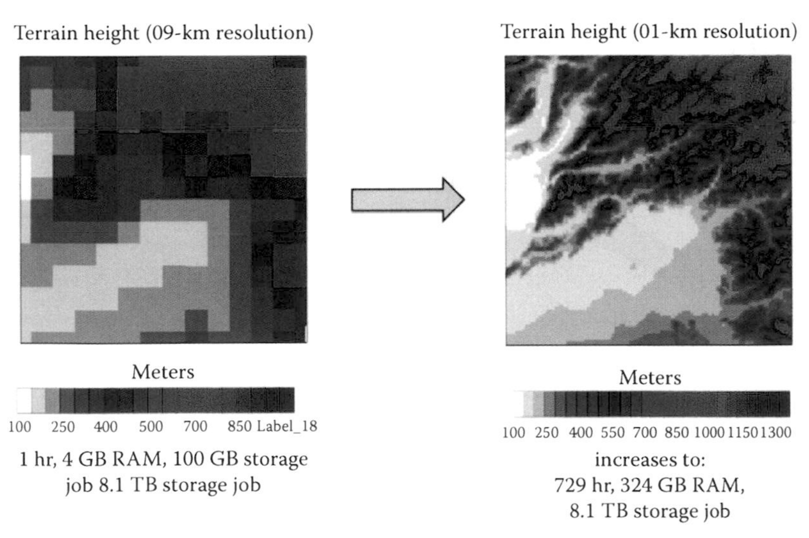

FIGURE 8.9 Example results from improving resolution from 9 to 1 km.

FIGURE 12.6 Static numerical model of a human heart.

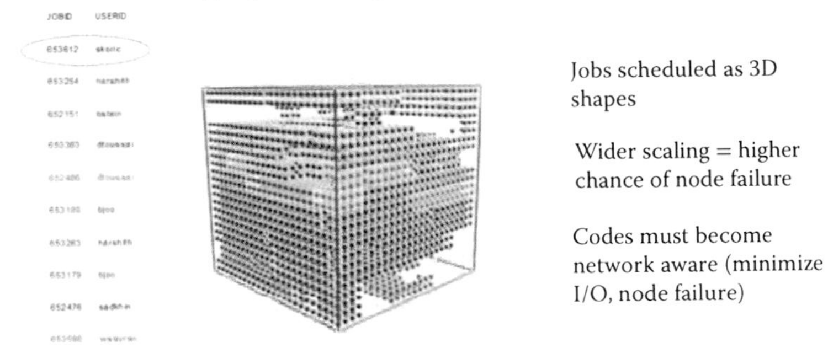

FIGURE 12.9 Graphical depiction of node utilization on Blue Waters, and how network and topological awareness is required to optimize applications.

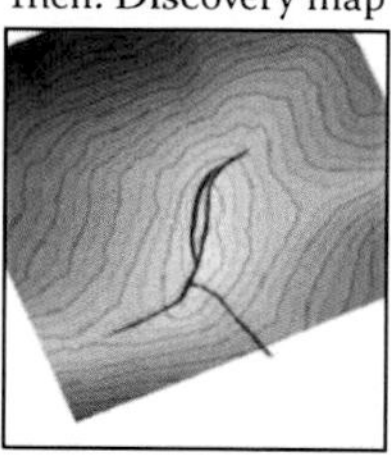
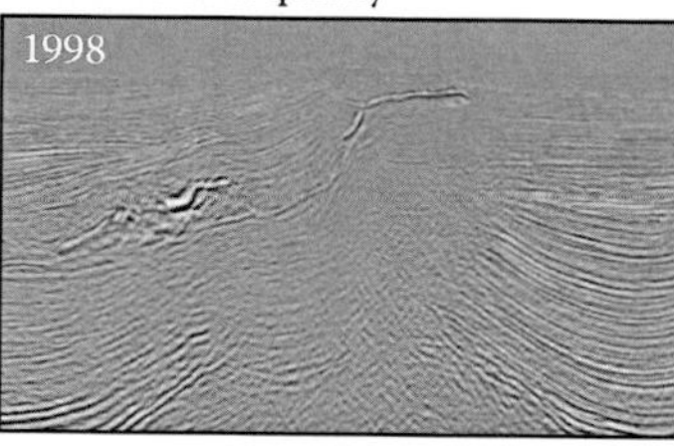
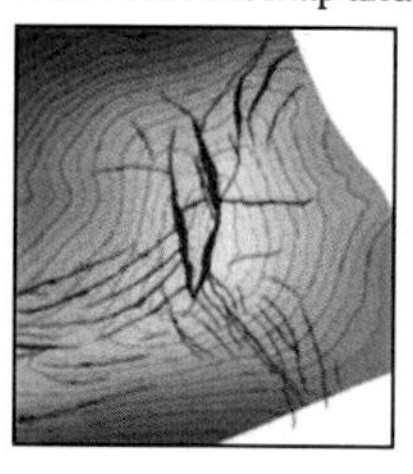
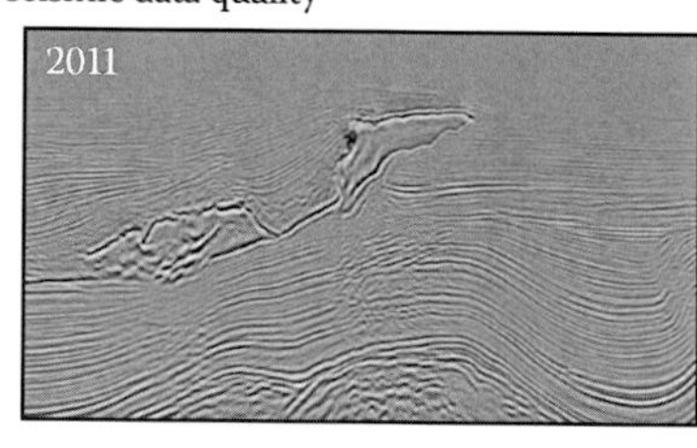

FIGURE 15.1 Improvements in seismic imaging from 1998 to 2011.

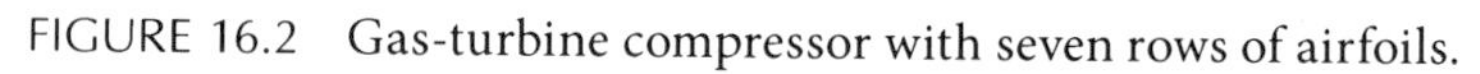

FIGURE 16.2 Gas-turbine compressor with seven rows of airfoils.

FIGURE 19.1 Model of "blades" in the low-pressure turbine.

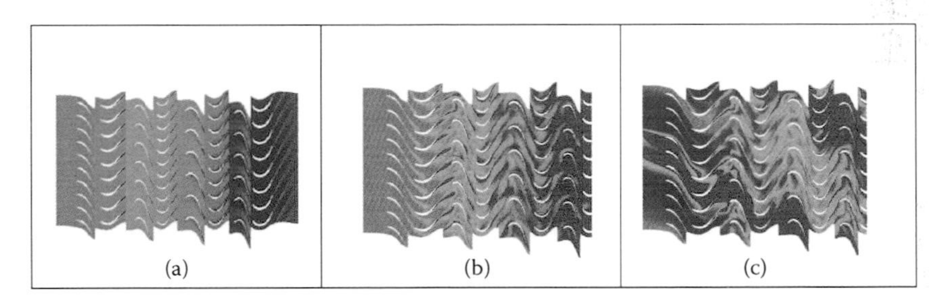

FIGURE 19.2 Entropy simulation through multistage low-pressure turbine. (a) Steady uniform inlet. (b) Unsteady uniform inlet. (c) Unsteady including strut wake. The white crescents are cross-sections of the blades. The simulation looks at the flow between blades in what is referred to as a "blade passage."

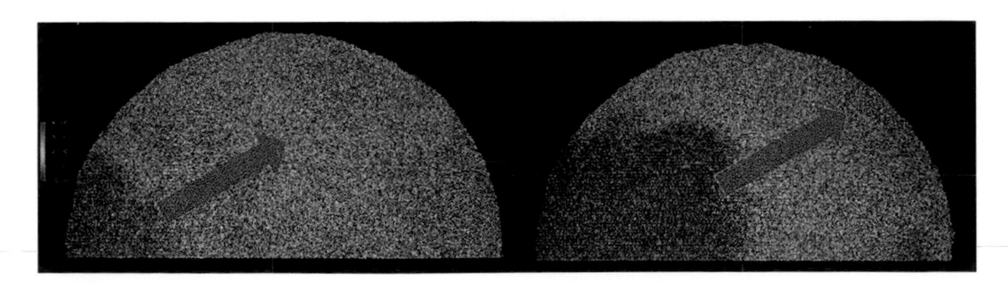

FIGURE 19.4 Critical nucleus propagating through droplet, where droplet mobility is denoted by color (blue: low mobility, white: average mobility, red: high mobility). As liquid water turns to ice, latent heat is released, causing the molecules surrounding the nucleus to heat up.

FIGURE 23.2 Lifeboat impact (1/10/100 millions of particles) application example of SPH-Flow.

CHAPTER 13

Industrial Letters of Testimony to the U.S. National Science Foundation

Merle Giles

CONTENTS

13.1 INTRODUCTION

The following excerpts are taken from the archives at the National Center for Supercomputing Applications (NCSA) at the University of Illinois at Urbana-Champaign. They are letters submitted by NCSA industrial partners to the Task Force on the Future of the NSF Supercomputer Centers Program. The NSF-supported Supercomputer Centers played a major role in U.S.

advancement of science and engineering research, and the Task Force Report was prepared as the Centers Program approached its ten-year anniversary.

These letters represent a rare compendium of industrial value that comes from supercomputing.[1] Do note the comments that state value in terms other than economic.

All content is in the form of quotes from the named authors, answering the following questions:

1. Does the existence of the Centers benefit your organization either as a resource for technical help and development, or by providing access to very high-end computational resources? If so, how?

2. Are there any quantitative measures?

NSF committee members were

1. Edward Hayes (Chairman), The Ohio State University
2. Arden Bement Jr., Purdue University
3. John Hennessy, Stanford University
4. John Ingram, Schlumberger, Austin, Texas
5. Peter Kollman, University of California, San Francisco
6. Mary Vernon, University of Wisconsin
7. Andrew White Jr., Los Alamos National Laboratory
8. William Wulf, University of Virginia
9. Nathaniel Pitts, Paul Young and Robert Voigt, NSF

13.2 ELI LILLY AND COMPANY

R.F. Abdulla, PhD, Director, Information Services (Scientific), Eli Lilly and Company, Indianapolis, Indiana

We are currently partners at the NCSA at the University of Illinois at Urbana-Champaign and are in the final stages of negotiation for our eighth consecutive year of membership. In general we have gained at Lilly through this partnership in the following areas:

- High Performance Computing Applications in Drug Discovery
- The dissemination of decision support algorithms

- Computational fluid dynamics and manufacturing plant modeling
- Scientific visualization and Virtual Environments
- The institutionalization of structure-based drug design
- The institutionalization of molecular bioinformatics
- The creation of an expert systems group within the HPCC discipline in Scientific Information Systems

The attempt to "quantify" the impact above, via outdated economic precepts and tools, does both the analysis as well as the disciplines represented above a disservice. I would state that the documented evidence suggests that the technologies represented above have added unique value to this and other companies. Much of the added value remains in our proprietary domains but a search of the literature will easily identify the added value we, at Lilly, have chosen to share.

It is my opinion that the paradigm of high-performance computing has changed since the creation of the NSF Supercomputing Centers in 1986. The drive to produce faster chips … has met with both cost and physics limits. Recent evidence … has suggested the dawn of a "new" paradigm in … parallel computing. The literature indicates, however, that in Europe, widely conceded as a theater with second tier chip design capability when compared to Japan and American (sic), this "new" high-performance computing paradigm was conceived of necessity in the 1980s. I refer, of course, to parallel computing in all its manifestations ranging from tightly coupled printed circuit boards and domain decomposition algorithms running across those boards, to the more sophisticated machines being built in the United States of various configurations.[2] Technically, these systems will be capable of running a few optimized algorithms at teraflop speeds before the end of the decade. Algorithm development is lagging seriously, and will continue to be a seriously limited factor in our capability to optimally utilize these systems.

If the preceding paragraph encapsulates a problem, it also defines a solution which could be invoked to solve the problem. The solution is one which we continue to follow in America, the successful model of the tripartite collaboration between the National Government, the Universities and the Industrial Corporations represented by the NSF Centers. It is evident that the mechanism for this to occur in the United States exists in the form of these National Supercomputer Centers. It is difficult to see such a collaboratorium exist at other than a National level. Other high-performance computing and communications centers may indeed spring up at the regional and state levels ... these in my view have been useful for educating our future technologists and technocrats, but can hardly be expected to play the role that has been played by the exemplary programs of the National Center for Supercomputing Applications at the University of Illinois. As a taxpayer, a technologist, and a partner at this center, I can state unequivocally that it would be a serious mistake to curtail this program.

13.3 AT&T

Marilyn Cade, Director, Technology and Infrastructure, AT&T Government Affairs, Washington, DC

AT&T participates as an Industrial Partner at the NCSA Supercomputing Center and via extensive high performance communications (gigabit) testbed supported through NCSA. We have been working with NCSA for several years in the critical areas of advancing high performance communications and computing in order to help to drive the emergence of these critical technologies.

AT&T has a strongly held and well publicized view that the private sector must, and is leading the way in building the information infrastructure—we, however, strongly support the concept of collaborative efforts in pre-commercial areas of research. The NSF Supercomputing Centers fit that model. The Industrial Partners program, for instance, is helping to provide the vehicle for new business users to participate in research endeavors pushing the envelop (sic), so to speak, in the capabilities of very early research prototypes of some of the technologies. This has brought significant benefit to the early testing and research we have undertaken in gigabit networking in conjunction with NCSA.

The Centers have also offered an environment which is helping commercial business experiment with emerging technologies—we see this as critical to helping to speed the awareness of new capabilities and to helping to create awareness of user needs for new yet untried applications of the high performance technologies supported through NCSA. The experimentation undertaken at NCSA and other centers is helping to develop "modeling" of new applications which can be focused at solving existing challenges in science and engineering.

13.4 CATERPILLAR INC.

Kem Ahlers, Manager of University Relations, Caterpillar Inc., Peoria, Illinois

Caterpillar has been an industrial partner at NCSA since December 1989. We joined by signing a multi-year multi-million dollar contract. We started by investigating the application of vector supercomputers to engineering analysis codes; then applying advanced visualization and virtual reality methods to product design; and most recently to finding ways to increase our competitive position by applying high-performance computing to information processing, business simulations, and related topics in our commercial area.

The advancement of computational science—including high performance computing—is a critical component in keeping American industry competitive. Increasingly our competitive position is determined by responding quickly to an ever-changing global marketplace. While no one can predict exactly what systems will be available ten years from now (another reason for our partnership with NCSA), we feel that they will be based upon distributed processing, high-speed networks and mass marketed microprocessors. We are depending upon ever-increasing performance at level or declining costs to meet our business needs.

I feel that the supercomputing centers and other University-based research institutions should be the focal point of the NII (National Information Infrastructure). NII research needs to be quickly assessable to American industry—and centers such as NCSA have demonstrated, through their unique business oriented Industrial Partner program, their ability to make evolving technology accessible. As an example, one of the benefits that we almost overlook is that Caterpillar's first exposure—and connection to—the Internet was via NCSA. This was very quickly used by many within our organization.

13.5 DOW CHEMICAL COMPANY

Randy Collard, Director of Communications and Information Technology Laboratory, Dow Chemical Company, Midland, Michigan

Developing new products not based on raw materials and/or low cost labor is critical to the future of the U.S. chemical industry. It is worth millions to it. The emergence of scalable parallel systems is limited by the development of effective software, (and we recommend a) shift from hardware to software with emphasis on joint efforts with both types (hardware and software) of vendors. Regarding benefits to DOW, NCSA's systems and expertise were critical in the development of our high performance computing capabilities. They have literally taught us to fish rather than just fishing for us. Their successful Industrial Partners program has made a successful technology transfer. This all-too-rare success has both saved Dow millions (of dollars) for existing operations and also helped us add millions more in new product development.

13.6 EASTMAN KODAK COMPANY

George Fisher, President & Chief Executive Officer, Eastman Kodak Company, Rochester, New York

Kodak has used the computer capabilities (at NCSA) in a number of effective ways. We simulated fluid flow inside coating hoppers of various shapes—something that could not have been done without the supercomputer. The knowledge we gained has been incorporated into our most advanced coating facilities in Kodak Park. Similar work was done by way of designing molds for injection molding processes.

Communications is a second, major thrust at NCSA and one that Kodak is only now tapping into. NCSA's web-browser MOSAIC enables imaging business opportunities, and Kodak plans to generate enabling tools to handle high quality images on the web. We are currently looking to see if a communication tool such as MOSAIC would be a viable system for internal data management, (including) how to integrate databases to facilitate parallel processes.

Kodak looks to NCSA to be at the front of evolving "supercomputing" technology, demonstrating the value and applications, and making the technology available to test-drive solving corporate problems. The application of such tools to make large amounts of information available is clearly a rapidly developing capability that has been driven by this center. The next critical focus is to use computational tools to improve how we take the huge amounts of data made available by the improved communications tools and convert it to knowledge. The recent focus on learning how to characterize and interpret large amounts of data using parallel database systems for commercial applications is such an example.

13.7 FMC CORPORATION

Ernest Plummer, Director of Computational and Analytical Sciences, Agricultural Chemical Group, FMC Corp., Princeton, New Jersey

Even within the purview of my own involvement with HPC it is difficult to provide specific quantitative evidence of the importance of HPC. The contributions are qualitative; they result in improvements in the way work is done and the value of the product derived. I envision a significant increase in simulations as substitutes for in-life environmental and toxicological testing. These will include environmental models as well as efficacy models for testing new pesticides and drugs. To be effective substitutes for living systems, these models will necessarily be complex. The complexity will lead to an increasing demand for HPC resources.

13.8 MORGAN GUARANTEE TRUST COMPANY OF NEW YORK

Charles Bonomo, Vice President, Morgan Guaranty Trust Company of New York (a subsidiary of J. P. Morgan & Co. Incorporated), New York, New York

J. P. Morgan (in 1992) was already an active participant in the area of high performance computing with three different parallel supercomputers being tested in-house. Our high performance computing needs are substantial and have been increasing.

The impact of distributed computing and real time financial data feeds in computer networks can be substantial in our work environment. We need very reliable computer networks, in addition to high speed links. We worked in conjunction with NCSA in a study of high speed network alternatives, e.g., ATM and HIPPI, and NCSA has provide us valuable advice.

We believe that the NSF Supercomputer Centers provided an essential bridge between academia and industry, have a major impact on the use of advanced computer technologies by U.S. corporations an greatly enhance the competitiveness in our business.

13.9 MOTOROLA

William Millon, Vice President and Director, Corporate Software Center, Motorola, Schaumburg, Illinois

We joined the NCSA Industrial Partners Program for the purpose of evaluating the degree to which high performance computing could be applied within the corporation. Making the supercomputer resources at NCSA available to our research engineers saved millions of dollars we otherwise would have been required to spend installing and operating our own supercomputer.

I believe that Supercomputer Centers Program is too narrow a name to apply to the work being done today at these centers. While I can only directly relate to NCSA, this center has evolved to much more than just supercomputing. The role of the centers should be broadened to encompass the more general term of high performance computing, and should include communication networks and information processing. The centers should play a key leadership role as places to adapt and nurture new technologies to help widen the evolving Information Superhighway.

13.10 PHILLIPS PETROLEUM COMPANY

Fred Lott, Director, Advanced Computing Center, Phillips Petroleum Company, Bartlesville, Oklahoma

High performance computing is a critical technology for the petroleum industry. In fact, the need for HPC continues to grow. In 1986, the primary driving force for HPC was seismic processing. Computational Chemistry was expected to use less than 1% of the supercomputer. Today, computational chemistry uses a significant portion of the Cray and technical cluster. In the near term, we hope to model refinery furnaces, so that we can meet emission standards. In the past, modeling furnaces has not been possible, but with the new computing power, furnace modeling is becoming feasible.

When Phillips purchased the Cray T3D supercomputer, we gained experience about the CM5/Thinking Machines supercomputer at NCSA/University of Illinois and the Cray/T3D at Pittsburgh Supercomputer Center. These NSF Supercomputer Centers purchase these supercomputers early in their technology cycle. These NSF centers are taking the technology risk rather than industry.

NCSA has played an important role in the technology transfer of engineering and computational science to Phillips Petroleum. The following are examples:

- Visualization capabilities for computational fluid dynamics.
- Computational chemistry code that gave Phillips insight into the structure of molecules.
- Network monitoring tools.
- Hierarchical Data Format (HDF)[3] that allows for data to be moved between different computer architectures, (including) to visualize the geology horizons. We cannot agree on the value, but we agree that it is an operating necessity.
- Phillips has been exposed to technologies that provide value to NCSA industry partners. For example, virtual reality was used by

Caterpillar Inc. to view the operation of heavy machinery. Phillips determined that, at this time, virtual reality did not provide value in our environment. The ability to determine that a technology does not add value saves mantime and cost.

- Study of Asynchronous Transfer Mode (ATM), a high-performance network technology.
- Phillips wants to benchmark against "best in class." We believe the NCSA industry partners are "best in class."

13.11 TRIBUNE INTERACTIVE NETWORK SERVICES

Eugene Quinn, General Manager, Tribune Interactive Network Services, Chicago, Illinois

The reason for our interest in NCSA is simple: we want to become more competitive. As a media company wrestling with the evolution from platform-dependent businesses such as publishing and broadcasting to non-platform-dependent businesses such as information, entertainment and transaction services, Tribune needs access to technology and research. NCSA's migration from application development for supercomputers to application development for meta-computing and broadband networks demonstrates the kind of pre-market innovating thinking that information industry leaders will find critical to serving customer needs in the future.

ENDNOTES

1. The full "Report of the Task Force on the Future of the NSF Supercomputer Centers Program, September 15, 1995, http://www.nsf.gov/pubs/1996/nsf9646/nsf9646.htm#_ftnref1
2. Intel's Hypercube, Kendall Square's system, Thinking Machines CM-5, and Cray Research Inc's Torus 3D architecture.
3. The HDF Group (www.thehdfgroup.com) spun out from NCSA as a non-profit corporation in 2005.

CHAPTER 14

A Livermore Perspective on the Value of Industrial Use of High-Performance Computing Resources at Lawrence Livermore, Los Alamos, and Sandia National Laboratories

Jeffrey P. Wolf and Dona L. Crawford

CONTENTS

14.1 INTRODUCTION

Lawrence Livermore, Los Alamos, and Sandia national laboratories, the scientific laboratories of the U.S. National Nuclear Security Administration's (NNSA) nuclear security enterprise, are perennially homes to some of the world's fastest supercomputers along with the broad capabilities needed to fully exploit them.[*,†] Over the years, several companies have entered into partnerships with these three laboratories (collectively known as TriLabs), whose critical resources support national security missions. Today, industrial partners are allowed access to unprecedented large and unclassified computing machines and the scientific expertise needed to help them accelerate research and development, solve high-impact business problems, and create competitive advantages. This chapter describes potential industrial engagement opportunities to help more companies understand how to engage in beneficial high-performance computing (HPC) and computational science projects with the TriLabs. Such engagements return value to the labs in the form of capability improvements, technology validations, and workforce invigoration stemming from productive interactions with industrial partners.

14.2 U.S. DEPARTMENT OF ENERGY AND NNSA: TOP INVESTORS IN CUTTING-EDGE HIGH-PERFORMANCE COMPUTER CAPABILITIES

For the past two decades, the Department of Energy (DOE) has been the largest sponsor of high-performance computing centers in the United States. Since the Top500 List rankings of the world's largest supercomputers were first compiled in 1993, high-performance computer systems at the TriLabs have occupied one or more of the top 10 positions.[1] Analyses of system data from the past decade of Top500 Lists indicate that the scale of

---

* This work was performed under the auspices of the U.S. Department of Energy by Lawrence Livermore National Laboratory under Contract DE-AC52-07NA27344.

computing used at the TriLabs is approximately 5 years ahead of that used by the reported leading industrial adopter, and 15 years ahead of companies limited to using only high-end workstations.

The history of computing at Los Alamos National Laboratory (LANL) can be traced back to its opening in 1943,[2] preceding the birth of the digital computer industry; at Sandia National Laboratories (SNL) to the early 1950s, when researchers performed predictive calculations of blast damage; and at Lawrence Livermore National Laboratory (LLNL) to its first year of operation in 1952. Each decade since has been punctuated by major system acquisitions and heroic usage of the largest, most powerful computer systems available for advancing science and developing solutions to the nation's toughest and most pressing challenges. This heritage continues today as expert TriLabs HPC users define the requirements for next-generation supercomputer architectures with the system software, productivity tools, algorithmic innovations, and application codes necessary to harness the ever-increasing scales and complexity of processing power, communications bandwidth, and data storage of the most advanced machines.

NNSA's demand for supersized machines is driven by national security concerns. TriLabs supercomputers perform important calculations that address needs in nuclear weapons, global security, and other national security areas that can affect people around the world. "Since 1992, the United States has observed the moratorium on underground nuclear testing [that has] ... required NNSA and its weapons laboratories to replace the functions of nuclear tests with ... nonnuclear experiments, highly accurate physics modeling, and improved computational power to simulate and predict nuclear weapon performance over a wide range of conditions and scenarios. This predictive power affords NNSA and the weapons laboratories the necessary tools to assess the stockpile, maintain its performance, continuously improve safety, respond to technological surprise and support future treaties."[3]

These comprehensive capabilities have been strategically developed through NNSA's Advanced Simulation and Computing (ASC) Program, which is aimed at maintaining high-fidelity, multiphysics predictive simulation capabilities and rigorous validation of experimental data. This computational infrastructure energizes the way thousands of scientists and engineers perform their work at the TriLabs, progressively reducing the time needed to simulate increasingly complex systems critical to national security.

14.3 TRILABS HPC CAPABILITIES PROVIDE OPPORTUNITIES FOR INDUSTRY

In an effort to address technical challenges of grave national and international concerns, the ASC Program spearheads a continuing campaign to produce the most realistic simulations possible using the most advanced HPC technologies, physics models, codes, and computer science. This pursuit drives engagements with the computer industry that push the capabilities of next-generation supercomputers forward so that realism can be extended to larger, more complex phenomena and systems.[4] This strategy has produced a large and unclassified superset of HPC capabilities at the TriLabs that can help produce breakthroughs for individual companies and entire industries.

For example, first-principles modeling and simulation at the quantum, atomic, and molecular scales can be used for predicting system behaviors. Scaled results from this ab initio approach can inform or be linked to higher level models and simulations at the mesoscale, continuum, or macroscale to achieve multiscale realism. With this capability, TriLabs scientists have successfully predicted wave action caused by sheer between two fluids using molecular dynamics simulations on 62 billion atoms (Kelvin–Helmholtz instability) (Figure 14.1). In industry, multiscale

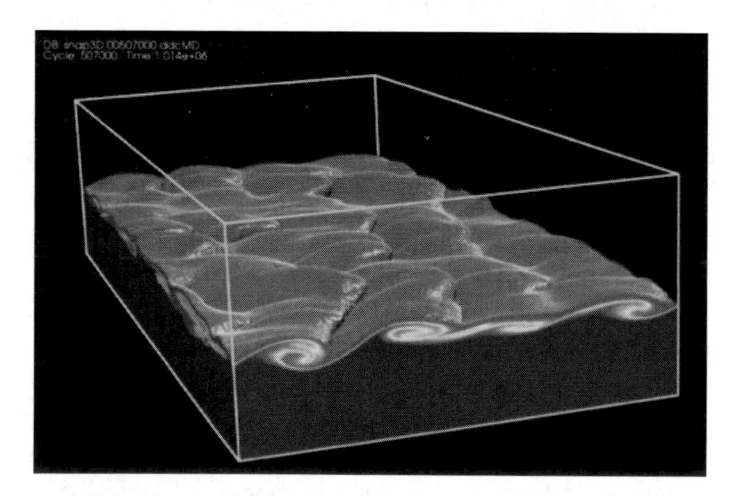

FIGURE 14.1 **(See color insert.)** TriLabs scientists predicted wave action caused by sheer between two fluids using molecular dynamics simulations on 62 billion atoms (Kelvin–Helmholtz instability). All of the processors on LLNL's BlueGene/L supercomputer ran for 6 weeks to produce this 1-cubic-micrometer simulation of liquid copper flowing like waves reaching a shore across liquid aluminum.

methods could be used to explore how molecular changes affect the properties of bulk materials over a large phase space. This approach is very different from the descriptive modeling approach typical in industry, where measured data is fitted to linear or exponential curves for interpolation between known data points. Descriptive modeling approaches are constrained by the limits of what can be measured and the ability to conduct controlled experiments. In many industrial applications, first-principles multiscale methods could be used to engineer new materials or optimize manufacturing processes more quickly and predictably than descriptive approaches.

In addition, multiphysics modeling and simulation codes can help researchers predict multiple mechanisms or properties of a system simultaneously—such as mechanics, fluid dynamics, thermal transport, chemical reactions, electromagnetic fields, or material properties. Many industrial products and manufacturing processes involve the mixing of solid, liquid, or vapor phases of materials, or the conversion of energy from one form to another. In such cases, multiphysics modeling and simulation can be used to accelerate process development and optimization; improve manufacturing yields; and reduce the time, cost, and number of prototyping cycles.

Highly scaled codes developed by TriLabs scientists and engineers can harness more than a million processors to execute one simulation. For example, groundbreaking heart simulations were developed and performed on LLNL's Sequoia supercomputer, an IBM BlueGene/Q system. During the months required to "shake down" Sequoia, while IBM and LLNL scientists installed and tested the machine in the process of bringing nodes online, LLNL made the system available for unclassified science calculations. In working on the Cardioid code for these heart simulations, IBM computational biologists contributed their expertise in cardiology, while LLNL scientists provided support in computational science, especially parallel algorithms (Figure 14.2). Commercial parallel simulation codes used in industry typically scale to tens or hundreds of processors, with leading codes capable of scaling to tens of thousands of cores. Highly scalable code development at the TriLabs is enabled and supported by multidisciplinary experts who can accurately model the underlying science and material properties, develop efficient parallel algorithms, manage complex data, and visualize simulation results.

High-dimensionality simulations, another TriLabs capability, enable the comprehensive study of complex systems and processes.

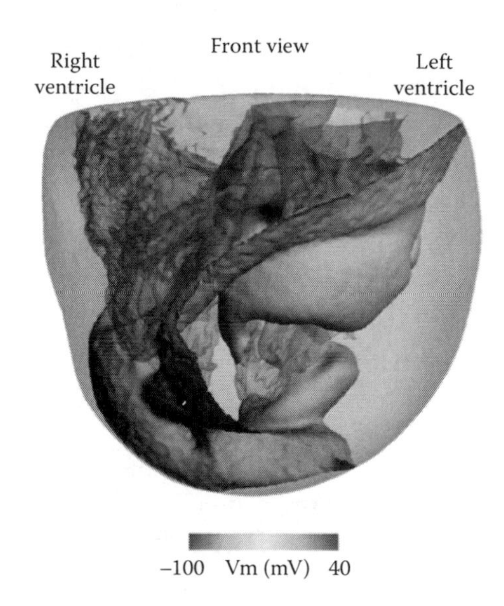

FIGURE 14.2 **(See color insert.)** The Cardioid code developed by a team of Livermore and IBM scientists divides the heart into a large number of manageable pieces, or subdomains. On Livermore's Sequoia supercomputer, the code ran near cellular-level resolution at 60 heartbeats per minute.

Although fully resolved, fully scaled three-dimensional (3D) simulations are not routine because of computer processing and memory limitations, they are performed when required. In industry, computing resource and time limitation cause many modeling and simulation users to forego 3D calculations. However, lower dimensionality simulation analyses can entirely miss valuable insights that would have been revealed with an added dimension or higher resolution. Through engagements with the TriLabs, companies can gain the knowledge to effectively perform, manage, and reap the benefits of more realistic 2D and 3D simulations.

Over the past two decades, the ASC Program has successfully transitioned its large software code base from each generation of supercomputer architecture to the next, and each transition has helped to further refine best practices for parallel software standards. This knowledge could greatly help companies grappling with a transition to parallel computing or to a new computing platform. Lessons will continue to be learned at the TriLabs with future transitions, making this area an ongoing engagement opportunity for industry.

14.4 CASE STUDIES OF TRILABS HPC TECHNOLOGY TRANSFER TO INDUSTRY

LLNL, LANL, and SNL are required by federal statutes to engage in technology transfer with industry, academia, and state or local governments. The Stevenson–Wydler Technology Innovation Act and the Bayh–Dole Act (Patent and Trademark Law Amendments Act), both enacted in 1980, along with the Federal Technology Transfer Act of 1986 and subsequent revisions to these landmark pieces of legislation, make technology transfer an integral part of each laboratory's mission. Although the technology transferred from the TriLabs to industry in the past has been predominately through licensing patents and copyrights, sponsored research, and open-source software releases, many HPC engagements with industry have been successfully carried out through several other mechanisms.

For example, the Cooperative Research and Development Agreement (CRADA), established by Stevenson–Wydler, provides the contractual framework for one or more laboratories to engage with one or more companies in a joint research project that can include industrial access to HPC systems. During the 1990s, several thousand collaborations between government and industry were carried out using CRADAs, with nearly 2000 involving DOE laboratories.[5] The federal government contributed technology-transfer funding to some of these engagements, and industry responded with enthusiasm. The High-Performance Parallel Processing Project (H4P) is a case in point where the federal government shared 50% of the project cost with industry in a $66 million program to accelerate the development of parallel computing.[6] The Industrial Computing Initiative (ICI), the industry-facing piece of H4P, used 11 CRADAs to structure the engagements between LLNL, LANL, Cray Research, and Thinking Machines. Many other companies and research organizations were also involved in this program, including Alcoa, Areté, AT&T, Boeing, Halliburton, Hughes Aircraft, and Xerox.[7] The 3-year H4P–ICI resulted in the development of many high-performance parallel software codes beneficial to DOE that were transformed into commercial applications for use by U.S. industry,[8] many of which are still in use in 2014.

ICI demonstrated that the federal government can successfully stimulate productive collaborations between national laboratories, supercomputer manufacturers, academia, and companies in the private sector by sharing the cost of engagements. In early 2014, DOE labs reported about 1000 active CRADAs, underscoring the acknowledged value of these collaborative public–private engagements between national laboratories and U.S.-based companies.[9]

Some TriLabs CRADA engagements that began in the 1990s continue as ongoing partnerships today. LANL engaged during this time with Procter & Gamble (P&G), and later with Boeing, Chevron, EMC, and Decision Sciences, with some of this work being HPC-related.[10] The Chevron–LANL relationship, described by Chevron as an alliance, is in its second decade of annual activity.[11] In the mutually beneficial CRADA with P&G, LANL scientists obtained real-world data to validate and verify statistical models, while P&G received advanced simulations of their high-volume, consumer-product manufacturing lines on which to base maintenance decisions. "Using this new capability, P&G reported that it was able to increase plant productivity by 44 percent, cut controllable costs by more than 30 percent, improve equipment reliability between 30 and 40 percent and reduce line changeover time from hours to minutes. It has been reported to have saved P&G more than a billion dollars in manufacturing costs."[12] The 27-year history of the P&G engagement with LANL is a marquee example of the business value that a sustained collaboration involving scientists and HPC resources at the TriLabs can deliver to companies with vision and well-specified problems to solve.

Sandia's industrial HPC engagements have also produced successes and well-publicized results. First and most notable has been its simulation-based engineering design partnership with Goodyear, which began in 1993. Although Goodyear's engineering management had already decided to move to HPC-based design, they needed the help of Sandia computational engineers to execute the workflow transformation. The resulting innovations from this engagement with Sandia saved Goodyear from being driven out of business by foreign competitors and revitalized the company. Loren K. Miller, who initiated and led this engagement at Goodyear, stated that without DOE's technology-transfer funding, this engagement would not have happened.[13] Another more recent example is Sandia's industrial HPC engagement with SunPower. Again, with the help of DOE funding,[14] Sandia assisted with developing and validating SunPower's HPC models and simulation software to predict and analyze photovoltaic generation based on installation and location. The resulting software, which scales up for grid-level analyses and down to a single processor for individual installations, is now widely distributed and used for solar-power installations around the world.[15]

In 1997, the EUV (Extreme Ultra-Violet Light) CRADA brought together three national laboratories, LLNL, SNL, and Lawrence Berkeley National Laboratory, and an industry consortium owned by Intel, Motorola, and AMD to develop commercially feasible next-generation lithography technology for the semiconductor industry. Other members of the consortium included Micron and IBM. This CRADA was one of the largest ever with a $250 million budget over 3 years.[16] Although HPC technologies were not salient elements of the work, this EUV engagement demonstrates the adaptability of the CRADA construct for companies and industries with innovation ambitions.

Work for Others (WFO) agreements have been more widely used than CRADAs for external TriLabs research and development engagements. In WFO engagements, one company specifies and funds one laboratory to perform company-directed turnkey research or, alternatively, a company is allowed access to laboratory assets to perform its own work. Companies seeking a laboratory to perform an HPC-enabled research project may prefer this approach, but it does not allow for active collaboration and the synergies that might result nor readily enable the transfer of computational science know-how. Although WFO engagements do not typically produce inventions, the results can deliver valuable insights and understanding that lead to innovations for sponsoring companies.

Procurements of next-generation HPC systems and technology have also driven industrial engagements across the DOE complex, because they often involve research and development initiatives to advance or commercialize early-stage HPC technologies. The NNSA's ASC Program has a history of investing in critical technology development and now considers collaborations with industry a vital part of its strategy. For example, the ASC PathForward and Industrial Partner Program funded the acceleration of InfiniBand software stack developments in 2004[17] and scalable Lustre and General Parallel File System development in 2008. In 2012, ASC's follow-on FastForward Program, jointly funded by DOE's Office of Science, Advanced Scientific Computing Research (ASCR), awarded five leading suppliers (AMD, Intel, IBM, Nvidia, and Whamcloud) $62 million for new technology development of extreme-scale computing.[18] In addition, ASCR provides $10–15 million annually in grants for Small Business Innovation Research and Small Business Technology Transfer. Many of these competitive awards have involved the transfer and commercialization of HPC technologies and scalable software developed at the TriLabs, producing opportunities for interaction between inventors and entrepreneurial grantees.

Jointly funded and open-source collaborations and technology transfers between the TriLabs and the HPC industry help move both technology and system scaling forward for the benefit of the broader HPC marketplace. For example, the Hyperion HPC test bed, which began in 2008 and is ongoing in 2014, teams 10 HPC ecosystem partners with LLNL to fund, build, and share access to a large Linux cluster sited at LLNL for testing next-generation HPC technologies at scale. This cluster is a resource that no one supplier would have procured alone, but access became economically attractive through an innovative cost-sharing approach.

External suppliers of operating system software, libraries, and development tools have also benefitted from interactions with the TriLabs, receiving valuable feedback from "power users," who routinely push the limits of scaling far ahead of the rest of the marketplace.

Finally, TriLabs computational scientists have been founders or active participants in many open-source communities, such as Open MPI, OpenMP, Open Scalable File Systems, Inc. (OpenSFS), and the Earth Systems Grid Federation. Many software codes developed at the TriLabs have been shared as open-source software to the HPC community at large, including operating system stacks, simulators, programming tools, code frameworks, visualization tools, math libraries, and compilers. HPC software developed by the TriLabs has given birth or a boost to many new companies, such as SchedMD (Slurm workload management software from LLNL), Livermore Software Technology Corporation (LS-DYNA finite-element simulation code from LLNL) and Kitware, Inc. (ParaView visualization and data-analysis software from LANL).

14.5 TRILABS INITIATIVES TO EXPAND INDUSTRIAL HPC ENGAGEMENTS

The TriLabs understand the value of more openly promoting HPC engagements with industry and providing the organizational focus needed for developing these engagements. In general, engagements that have occurred in the past have been the result of persistent companies with specific needs that could be met at the TriLabs. In many cases, these companies made some fortuitous connections to the interests and skills of TriLabs personnel who, in turn, stepped up to make these engagements happen and bear fruitful results. More systematic educational and developmental outreach to potential supercomputer users by national laboratories and academia is needed to expand the HPC market, along with more outreach to commercial modeling and simulation software suppliers to help and encourage them to scale their codes. The Council on Competitiveness[19]—a group of university presidents, corporate CEOs, labor leaders, and national laboratory directors working to ensure U.S. prosperity—has identified HPC as a key technology for boosting American productivity and economic prosperity and for enabling American manufacturing companies to compete effectively in global markets. The council's studies make it clear that more needs to be done to promote and provide access to the benefits of HPC technologies for companies that may be uninformed and missing business growth opportunities.

Analyses of the CRADA-enabled and other industrial HPC engagements with the TriLabs yield some important observations that are common to the outreach and engagement histories of most federally funded supercomputing centers. Most of the participating companies came in with prior experience and predetermined intent to use HPC, and they proactively sought principal investigators or procurement decision makers at the TriLabs to initiate these engagements. Most, but not all, of the nonprocurement engagements have been with large, profitable companies with sizable research and development budgets. Rare on the list are participants with no previous use of HPC, who were introduced and educated by TriLabs' personnel about supercomputing benefits and subsequently became committed adopters of HPC technologies. Small businesses that would have received federal government preferences in the TriLabs' CRADA selection processes are also difficult to find among the participants in these early CRADA engagements, except for some very focused early adopters.

Although the TriLabs have been successful at bringing the HPC supply chain forward, this effort has not been sufficient to broaden the adoption of HPC technologies across the U.S. industrial base. HPC suppliers do offer products and services to the market, but the total price of entry has been higher than most companies would pay to explore the possibilities. The range of necessary expertise needed to implement a full HPC solution usually exceeds what is available within a single customer organization that has not already adopted HPC. This full range of expertise is also absent in all but a few HPC suppliers, and those with this capability value it as a competitive advantage to be offered to others only at a high price.

In 2009, LLNL, SNL, and NNSA took proactive measures to increase industrial engagements by creating the Livermore Valley Open Campus (LVOC), with LLNL's first investment specifically focused on HPC outreach. The HPC Innovation Center at LVOC opened in mid-2011 with the goal of engaging companies to develop, prove, and deploy HPC solutions that solve high-impact problems and create competitive advantages. The center offers industrial partners access to a powerful combination of computing resources and computational expertise from across LLNL and its HPC ecosystem partners.

At the HPC Innovation Center, LLNL's HPC assets are presented for the first time as a primary attraction for industrial engagements, supported by broad scientific and engineering expertise and unique experimental facilities. Companies have been responding with interest in exploring engagement opportunities. IBM Research and the United Kingdom's Hartree Center at STFC-Daresbury have established formal collaborations with the HPC Innovation Center to broaden the range of subject-matter expertise and geographical reach for multicompany and multi-institutional engagements. The Catalyst system, a Cray CS300 cluster supercomputer, is the latest system to be added to the center. The machine is cost-shared by Cray, Intel, and LLNL and used for external big data HPC research projects that can benefit from access to more than 300 terabytes of combined dynamic and nonvolatile memory per processor. Industrial engagements to-date include several oil and gas companies seeking access to LLNL's geomechanics expertise and predictive simulation codes; industrial manufacturers seeking for the first time to include HPC in their product and system design processes; electric power utilities analyzing future grid operating scenarios; and defense contractors developing new technologies and complex systems. Among these engagements are several in which the industrial partner had not been committed to HPC use before the start of engagement discussions.

Also in 2011, in an effort to expand its HPC outreach, LLNL sponsored and conducted a pilot program for energy-related projects called the "hpc4energy" incubator. From 30 applicant proposals, 6 were selected for short-term, no-cost, publishable energy-related collaborations, featuring access to LLNL's HPC systems and its computational and domain scientists. The award recipients were GE Energy, GE Global Research, ISO New England, Potter Drilling, Robert Bosch, and United Technologies. The application areas for these projects included carbon capture, utilization, and sequestration; energy-efficient buildings; liquid-fuels combustion; and smart grid, power storage, and renewable energy integration. The projects concluded in 2013 with each participating team having achieved substantial reductions in simulation execution time for their calculations of interest through increased parallelization and the use of large supercomputers. Participants came away with volumes of resulting data and new insights. Because the results and gained insights are applied to new product design or inserted into ongoing business operations, the benefits for the participants and their target markets are not yet known. Realization of business impacts from HPC projects such as these take an average of 2 years.[20] From LLNL's perspective, the hpc4energy incubator project execution went well, but resulting industrial HPC adoption and energy technology impacts from its investment are as yet unclear.

In addition, periodic workshops have brought together DOE leaders and industry for open dialogue on industry challenges and learning. For example, the 2012 Grand Challenges Workshop on Advanced Computing for Energy Innovation built on the momentum of three previous workshops to produce four recommendations on ways DOE's advanced computing resources can impact the energy future of the United States:

1. Improve the usability and availability of DOE-developed advanced computing solutions
2. Engage the independent software vendor community to promote energy innovation
3. Implement policies to facilitate adoption of advanced computing for energy innovation
4. Establish an Advanced Computing for Energy Program within DOE[21]

14.6 A BRIGHT FUTURE

HPC technologies, effectively applied, offer companies in a wide range of industries the potential to innovate, transform their businesses, and improve their competitiveness in the global marketplace. Published examples demonstrate that companies with vision, commitment, and persistence can realize this potential in many business activities such as research and technology development; engineering; manufacturing; logistics; and big data analytics for marketing, sales, and operations management.[22]

Efforts are under way at each of the TriLabs to make themselves more accessible and business friendly within their operational constraints established by statutes, federal charters, DOE policies, and their respective mission priorities and governance. The TriLabs have primary national security missions and are, at best, secondarily organized to create and execute large numbers of engagements with industry; however, they are motivated to transfer technology to industry, using well-established contractual vehicles that have been successfully implemented for decades. Companies that proactively seek opportunities to initiate contact and engage with the laboratories, and that persist in defining relevant projects and scopes of work that can positively impact their businesses, can receive highly valuable results from engagements with the TriLabs.

The returns on investment and gains in knowledge can be material and transformational to engaged businesses as a result of affordable access to: substantial supercomputers for developing and testing HPC capabilities; highly scalable software codes; formidable big data capabilities; and a broad population of leading domain science, engineering, and computational expertise. Results can include accelerated development of breakthrough products and technology; optimization of materials and manufacturing processes; timely solutions to complex high-impact problems; and substantial reductions in business risk, operational costs, and cycle times.

The time is right for expanding HPC-capability engagements between the TriLabs and American industry for the benefit of the nation. In today's world of scarce parallel computing expertise, the TriLabs shine as gleaming repositories of valuable capability and knowledge that could be applied to industrial needs. The challenge for companies and political leaders is to bring these national resources to bear, in a timely fashion, on the most relevant and impactful needs at hand.

REFERENCES

1. TOP500® Supercomputer Sites website, http://www.top500.org (accessed December 3, 2014).
2. Alan B. Carr, "The History of Computing from Punched Cards to Petaflops," *National Security Science*, Los Alamos National Laboratory, Los Alamos, NM (April 2013).
3. National Nuclear Security Administration's Research, Development, Test, and Evaluation website, http://nnsa.energy.gov/aboutus/ourprograms/defenseprograms/stockpilestewardship (accessed March 23, 2014).
4. James A. Ang, Paul J. Henning, Thuc T. Hoang, and Rob Neely, *Advanced Simulation and Computing: Computing Strategy*, SAND 2013-3951P (Washington, DC: Office of Advanced Simulation and Computing, NNSA Defense Programs, 2013).
5. David C. Mowery, *Using Cooperative Research and Development Agreements as S&T Indicators: What Do We Have and What Would We Like?*, proceedings from Strategic Research Partnerships: a National Science Foundation Workshop (Arlington, VA: NSF, 2003).

6. Michel G. McCoy, "The Industrial Computing Initiative," *Energy and Technology Review*. UCRL-52000-94-10 (Livermore, CA: Lawrence Livermore National Laboratory, 1994).
7. Alice Koniges, *High Performance Parallel Processing Project/Industrial Computing Initiative: Progress Reports for Fiscal Year 1995*, UCRL-ID-123246 (Livermore, CA: Lawrence Livermore National Laboratory, 1996).
8. Alice Evelyn Koniges (Ed.), *Industrial Strength Parallel Computing* (San Francisco: Morgan Kaufmann, 2000).
9. National Science Foundation's National Center for Science and Engineering Statistics website, http://www.nsf.gov/statistics.
10. Los Alamos National Laboratory website, http://www.lanl.gov.
11. Chevron Corporation website, http://www.chevron.com.
12. National Institute of Standards and Technology, *Federal Laboratory Technology Transfer: Fiscal Year 2009* (Washington, DC: NIST, U.S. Department of Commerce, 2011).
13. Loren K. Miller, "Simulation-Based Engineering for Industrial Competitive Advantage," *Computing in Science and Engineering* **12**(3) (IEEE Computer Society, May/June 2010), 14–21.
14. Lara Getz, *CRADAs: Cooperative Research and Development Agreements* (Washington, DC: American Security Project, 2011).
15. Tim Townsend and John Wilson, *SunPower Yield Report* (San Ramon, CA: BEW Engineering, 2013).
16. David C. Mowery, "Using Cooperative Research and Development Agreements as S&T Indicators: What Do We Have and What Would We Like?," in *Strategic Research Partnerships: Proceedings from an NSF Workshop*, Arlington, VA (NSF 01-336) (August 2001).
17. Bob Meisner, "Project Provides Thirteen Years of Industry Shaping Deliverables," *ASCeNews,* Lawrence Livermore National Laboratory, Livermore, CA (September 2008).
18. Lawrence Livermore National Laboratory, Public Affairs Office, *DOE Collaboration to Accelerate High Performance Computing*, NR-04-11-03 (November 2004).
19. Council on Competitiveness website, http://www.compete.org/about-us/initiatives/hpc (accessed December 3, 2014).
20. Earl C. Joseph, Steve Conway, and Chirag Dekate, *Creating Economic Models Showing the Relationship Between Investments in HPC and the Resulting Financial ROI and Innovation—and How It Can Impact a Nation's Competitiveness and Innovation,* No. 243296, Vol. 1 (Framingham, MA: IDC, 2013).
21. Workshop on Grand Challenges in Advanced Computing for Energy Innovation website, http://ams.labworks.org/challenges_workshop (accessed December 3, 2014).
22. Council on Competitiveness website, http://www.compete.org/publications (accessed December 3, 2014).

II

Case Studies

CHAPTER 15

High-Performance Computing Business Case in the Petroleum Sector

Keith Gray

CONTENTS

15.1 INTRODUCTION

As discussed in Chapter 1, the petroleum sector represented the most dominant industrial use of high-performance computing (HPC) in 1982 (10%). Since then, the use of HPC in oil and gas has continued to accelerate, driven by the requirements of seismic imaging. In a process known as reflection seismic prospecting, sound waves are sent into the earth and bounced off of rock layers, creating energy waves that are recorded as they return to the surface. This is the most accurate method available for understanding subsurface geological features, and critical to discovering and progressing resources. It enables researchers to understand the subsurface for prospect generation and resource appraisal before field development. And once the fields are developed, it enables developers to monitor oil reservoirs over time.

BP, formerly British Petroleum, is a British multinational company that is the product of several mergers. In 1987, BP acquired the U.S. Standard Oil Company of Ohio, or SOHIO. In 1998, it merged with

Amoco Corporation. BP then acquired ARCO (America's Atlantic Richfield Company) in 2000. In such a globally competitive industry, this consolidation was necessary to amass the resources required for productive operations. Importantly, these companies all brought a history of HPC-enabling business breakthroughs, and today BP has the world's largest commercial research supercomputing center.

The first drilling rig used to produce oil in the United States occurred in Titusville, Pennsylvania in 1858.[1] The drilling location was identified by oil seeps at the surface. Edwin Drake used a steam engine to power the drill rig at the rate of 3 ft per day, and it was drilled to a depth of 69 ft. The technical breakthrough that enabled this success was the use of cast iron pipes in 10-ft sections to prevent the hole from collapsing. But the prospects identified by surface seeps were quickly played out, and new technology was needed to identify deeper reserves.

Seismic imaging was first utilized in exploration in 1921 by a team of physicists and geologists (William Haseman, Clarence Karcher, Irving Perrine, and Daniel Ohern) working in south-central Oklahoma. The team blasted dynamite in shallow holes, and the sound waves were recorded by three seismographs located 300 m away. The seismographs contained a light source and a mirror that vibrated as the energy returned to the surface, and the light was recorded on a rolling photographic film strip. With each subsequent dynamite shot, the energy moved the seismographs along a line. The photographic recordings needed to be hand-corrected to create a vertical profile of the subsurface. The person who made these corrections was called the "computer." The details of the experiment and the compromises necessary to carry it out given the state of technology available at that time are described by William Dragoset in an article of The Leading Edge celebrating the 75th anniversary of the Society of Exploration Geophysicists.[2]

In the 1950s, seismic processing started to use real "computers." One of the first people to recognize the opportunities of using the computer in seismic exploration was Enders Robinson. An IEEE Oral History is presented as an interview of Dr. Robinson by Andrew Goldstein conducted in March 1997.[3] Robinson credits his discoveries on the opportunity to work as a graduate student under Norbert Wiener at America's Massachusetts Institute of Technology (MIT). The critical mass of brilliant people with mathematics and computing skills were credited with creating these breakthroughs. But Robinson's key discovery was recognizing the convolutional model of the earth—the seismic waves created when dynamite is exploded travel through the earth, and are reflected back to the surface when they hit rock layer interfaces. But these primary reflections were hidden by noise such as multiple reflections, surface waves, and refracted waves that hide geological information. Robinson proved that the mathematics of deconvolution could remove noise and improve the quality of the image. Oil companies recognized the value of these technologies and began to sponsor the Geophysical Analysis Group at MIT in 1953. This financial support attracted students into geophysics. Many of these students left to start careers in the oil and gas industry.

Breakthroughs in computing hardware and software were critical in sustaining the growth of seismic imaging. The release of Fortran (derived from FOrmula TRANslating System) by IBM in 1957 (see Chapter 1) eased the programming burden, allowing geophysical programmers to write in natural language and mathematical formulas instead of low-level machine assembly language. IBM mainframes dominated seismic processing through the 1970s into the late 1980s. Batch job scheduling systems, tape-to-tape processing (because hard drive storage systems were too small to hold seismic volumes), and specialized array processors like the IBM 3838 and FPS-190L were common. Specialized minicomputers were also used, from companies as varied as Texas Instruments, Interdata, Raytheon, and DEC. The oil and gas industry was one of the first commercial users of Cray vector systems. Massive parallel systems like the Thinking Machines CM-2 and CM-5 were used in the late 1980s and 1990s. SGI systems became very popular in the 1990s as the visualization systems grew large enough also to be used for processing; they undercut costs of Cray systems.

By 1999, BP and other oil companies were experimenting with Beowulf commodity clusters (see Chapter 1). In 2000, ARCO replaced a time-sharing agreement to use Cray supercomputers with a 32-node cluster. BP deployed a 256-node cluster that delivered 4 teraflops of computing power in 2002 using Hewlett Packard systems with Intel Itanium CPUs. These systems were chosen because they were able to support 32 gigabytes of memory per node; the large address space and fast memory bandwidth made software development more effective (large memory caches allow developers to focus on geophysics rather than subdividing problems across multiple computers).

The growth of clusters dramatically increased the demands for and pressure on parallel file systems. In 1999, BP HPC had a total storage capacity of 17 terabytes (TB), all directly connected to SGI Power Challenge and Origin systems. In a session at the 2003 Supercomputing conference, BP stated its aspiration to deploy a 10 TB file system with the ability to feed clusters at 1 gigabyte/s. BP wildly exceeded these hopes. By 2006, BP's computing facility had grown to 75 teraflops of computing power, with 10,000 CPUs and 1.8 petabytes (PB) of data storage. By 2013, BP had deployed multiple Lustre and Panasas file systems, the largest of which could hold 3.5 PB of storage and was capable of delivering more than 50 GB/s. In total, BP's parallel file systems can deliver more than 400 GB/s.

At BP's U.S. headquarters in Houston, the HPC support team now supports the largest commercial research computing facility in the world, boasting more than 2.2 petaflops (a technical abbreviation for 2.2 quadrillion floating point operations per second) of compute power on 100,000 CPUs, alongside 24 PB of data storage and 1 PB of main memory.

Because of the size and scale of its HPC systems, facility limitations became a critical issue for BP by the mid-2000s and challenged BP's ability to continue growing HPC. By 2004, power demands exceeded the available 1.0 megawatt uninterruptable power supply (UPS) and diesel generator backup capabilities. And it did so in a relatively spectacular manner. When new SGI Altix systems were powered on, the main breaker tripped, leaving the building that housed the HPC and 3500 people on campus without power. Additional investments in UPS and cooling systems allowed the space originally designed for IBM mainframes to continue operating, but a new facility was clearly needed. That new facility, which is devoted entirely to HPC, opened in 2013, supporting 2.5 megawatts of peak power draw. At 110,000 ft,[2] the facility now has the ability to grow to support 9 megawatts of power. The power utilization efficiency (PUE) of the new facility is designed to a goal of 1.35, a 30% efficiency improvement compared to the previous building.

15.2 BUSINESS VALUE ENABLED BY HIGH-PERFORMANCE COMPUTING

Seismic imaging improvements have been made possible by increasing three-dimensional (3D) imaging complexity using supercomputers. As reported in the World Census of Supercomputers in 1982,[4] HPC was used at ARCO, SOHIO, Exxon, Chevron, Texaco, and Shell (NL/UK). BP's supercomputing timeline more precisely identifies the use of a CRAY-1 as early as 1976, followed by a CRAY-2 in 1985.

Exxon Production Research began work on a 3D seismic system in 1963. They acquired the first 3D seismic survey over the Friendswood field near Houston in 1967.

In 1972, Geophysical Service Inc. (often abbreviated GSI) conducted a research project for six partners (Chevron, Amoco, Texaco, Mobil,

Phillips, and Unocal) to evaluate 3D seismic imaging. The Bell Lake field in southeastern New Mexico was a structural play with nine producers and several dry holes. It also had sufficient borehole data to ensure that 3D seismic images could be correlated to subsurface geology. The acquisition phase took only about a month, but processing the half-million input traces required another 2 years. Nonetheless, the project was a defining event in seismic history because the resulting maps confirmed the field's nine producers, condemned its three dry holes, and revealed several new drilling locations in a mature field.

Development of 3D seismic imaging and analysis was one of the most important technological breakthroughs in this industry, where profitability is so closely tied to innovation and technology. At last, the subsurface could be depicted on a rectangular grid that provided the interpreter with detailed information about the full 3D subsurface volume. The images produced from 3D data provided clearer and more accurate information than those from 2D data.

The value of 3D seismic exploration was described in an article written by William Aylor of BP and presented at the 1999 Offshore Technology Conference. The presentation described the major turnaround in exploration performance in the 4-year period from 1992 to 1996 that can be attributed to the value of 3D seismic acquisition and processing. During this period, BP Amoco made major improvements in its various measures of success:

- Production replacement (the ratio of discovered oil to produced oil, an indicator of an oil company's long-term sustainability) improved from 60% in 1992 to 179% in 1996
- The cost of finding oil dropped from $8 per oil barrel (bbl) in 1991 to under $1 per barrel in 1996[5]
- The new resources' discovery improved from 200 million barrels of oil equivalent in 1991 to 1 billion barrels of oil equivalent in 1994–1996
- The exploration drilling success rate jumped from 13% in 1991 to 44% in 1996

Drilling efficiency is a key driver of production, particularly because a deepwater well in the Gulf of Mexico can cost over US$150 million. Innovation, accelerated by HPC, informs decision makers and lowers uncertainty for determining the placement of wells. The combination of computing technology, programming support, and seismic processing provides a competitive advantage in understanding, or seeing, geophysical formations that promise better placement of wells.

One example of a competitive advantage provided by research is BP's development and acquisition of the industry's first Wide Azimuth Towed Streamer (WATS) for subsalt imaging, deployed at the Mad Dog oil field in the Gulf of Mexico in 2005.

15.3 BP's BREAKTHROUGHS IN SEISMIC ACQUISITION AND WATS

In an article published in the Houston Chronicle on October 28, 2013, Zain Shauk interviewed John Etgen, the BP distinguished advisor for seismic imaging, who stated: "BP in 2004 pioneered the WATS method, a new approach to seismic that would change the industry. The process captures more sound waves with the same set of cables by using multiple boats firing air guns." "Each boat sends sound waves into the ground, delivering reflections to the surface from multiple angles. Powerful computers use mathematical processes captured in programmable algorithms to combine the echo images, creating a sharper view of what lies beneath the salt," Etgen said.

The effect is similar to sports broadcasts that use multiple cameras to show the action from different angles. If one angle produces a distorted or incomplete image of the underground rocks, measurements of other reflection angles help fill in the missing parts. "The more of that thing that I can catch and record, the better chance I have of making an image," Etgen said. "The wide azimuth survey produced sixteen times as much data as conventional seismic surveys," he said.

The "Mad Dog" field was discovered in the Gulf of Mexico in 1998, and improvements in wide azimuth seismic technology have significantly impacted image quality between then and 2011 (see Figure 15.1). The oil-producing structures at the top of the anticline were completely obscured by salt structures, requiring more complex imaging procedures. The first WATS field trial was conducted over the Mad Dog field. This first experiment would cost 300% more than conventional methods (a multimillion-dollar risk). To gain management approval for this experiment, the conventional narrow azimuth acquisition and the WATS acquisitions were simulated in the computer using complex algorithms. BP then processed the two synthetic datasets to compare the improvements. Only through BP's simulation, using HPC, would this risk be tolerable. For 6 months, BP dedicated over 50% of its computing resources to fast-track processing

Then: Discovery map and seismic data quality

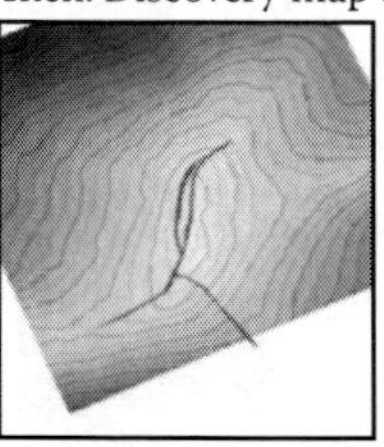
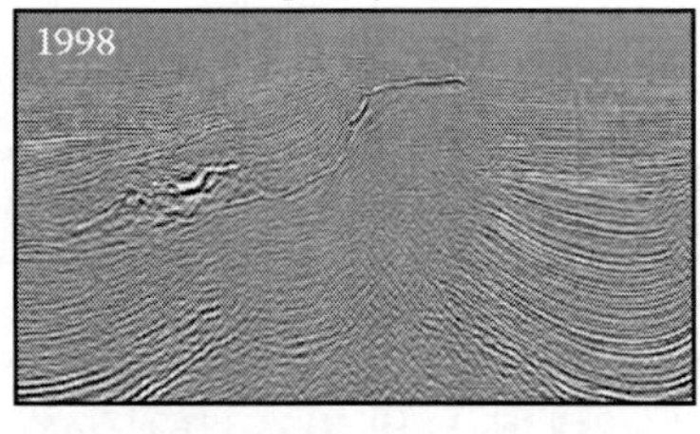

Now: Current map and seismic data quality

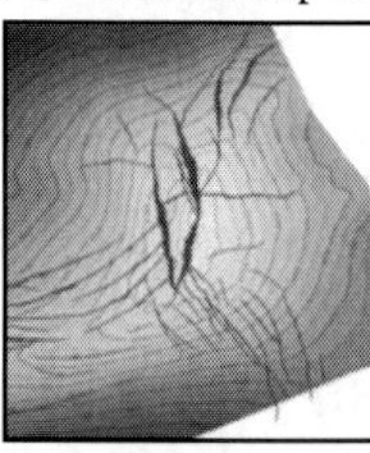
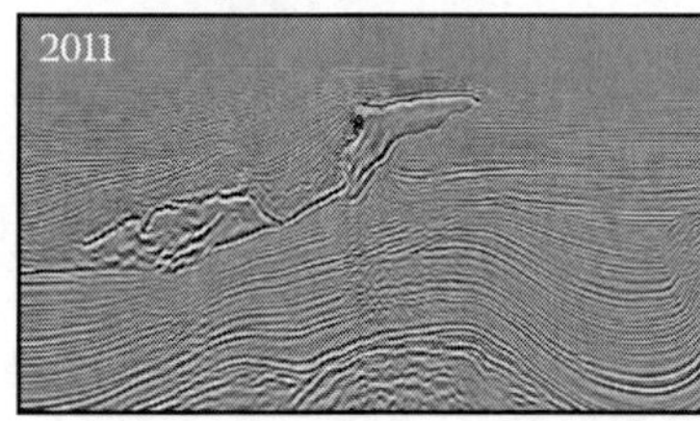

Example of data improvement over the Mad dog field:

BP developed and acquired the industry's first wide azimuth towed streamer (WATS) survey there in 2005.

Since then, we have continually improved the methodology and imaging.

FIGURE 15.1 **(See color insert.)** Improvements in seismic imaging from 1998 to 2011.

of the data as it came off the boat. By the time the acquisition was 50% complete, the image quality exceeded the conventionally acquired data. Because we could now "see" the geologic structures and faults, we could plan the development of deepwater subsalt reserves.

Collaboration is a common business model in high-cost projects such as this, so ownership of the field was shared by BP (60.5%), BHP Billiton (23.9%), and Chevron (15.6%), and operations were accomplished using a spar[6] oil platform hull that was manufactured in Finland.[7] The discovery was followed by drilling in 1999 to a depth of 22,410 ft. This all occurred despite the fact that Mad Dog was one of a number of discoveries occurring in rock folds covered by salt tongues, making seismic imaging difficult. These breakthroughs in imaging quality have been delivered through a long-term investment in seismic research and HPC that enables these breakthroughs.

Investments in seismic imaging research and HPC are also credited for delivering dramatic improvements in the cost-effectiveness of seismic acquisition in the period between 2008 and 2013.[8] Simultaneous source seismic acquisition has enabled breakthroughs. BP has publicly discussed the 80% reduction in cost and time for seismic surveys in North Africa and North America that has resulted from using simultaneous source technology. Historically, seismic acquisition required positioning an energy source (for example, dynamite or sound waves) at a single location, then waiting until the energy propagated to the subsurface target and back to the surface before energy was fired at another position. Through the clever use of compressive sensing technologies learned from other industries (like photography and medical imaging), BP was able to acquire data from multiple energy sources firing simultaneously. Because the recorded seismic data is all "blended" together, the distinct sources must be separated in the computer. This requires significant computing resources, which pay for themselves through lower costs and higher quality seismic acquisition.

Imaging R&D teamwork requires an HPC team with highly specialized skills in geophysics, computer science, and mathematics. The BP HPC Team has three distinct subgroups. First, the HPC *Development* Team works directly with researchers to develop application codes, optimize algorithms, and automate processes. It typically takes 3 to 5 years for a new PhD to develop the skills to be fully productive in this highly technical field. The HPC *Operations* Team then provides seismic processing and data management skills to enable researchers to focus on the delivery of new technology. Rounding out the effort is the HPC *Systems* Team, which manages the huge computing center, implementing leading-edge technology at scales that rival many of the supercomputers in federal labs and universities.

To identify and train these specialists, BP is a member of consortia at the University of Texas (TACC), University of Illinois (NCSA), and Rice University (Ken Kennedy Institute). We recruit three to four summer interns annually to identify students with the aptitude and interest in HPC.

CONCLUSIONS

As powerful as BP's 2013 supercomputing capacity is, an estimated >2× growth in computing power per year is needed to keep pace with imaging requirements. (See Figure 15.2) A presentation by Scott Morton (Hess), Henri Calandra (Total) and John Etgen (BP) describes the need for an exaflop supercomputer in oil and gas.[9] Demand from the industry to increase resolution (allowing us to see smaller structures and faulted rock structures) and understand rock properties (allowing us to predict if the rocks will bear oil and gas reserves) are constantly feeding the need for more complex and challenging algorithms and more powerful supercomputers.

We continue to test hardware acceleration technologies, such as graphics processing units (GPUs), but this decision is difficult. Increased software complexity through the use of accelerators needs a clear

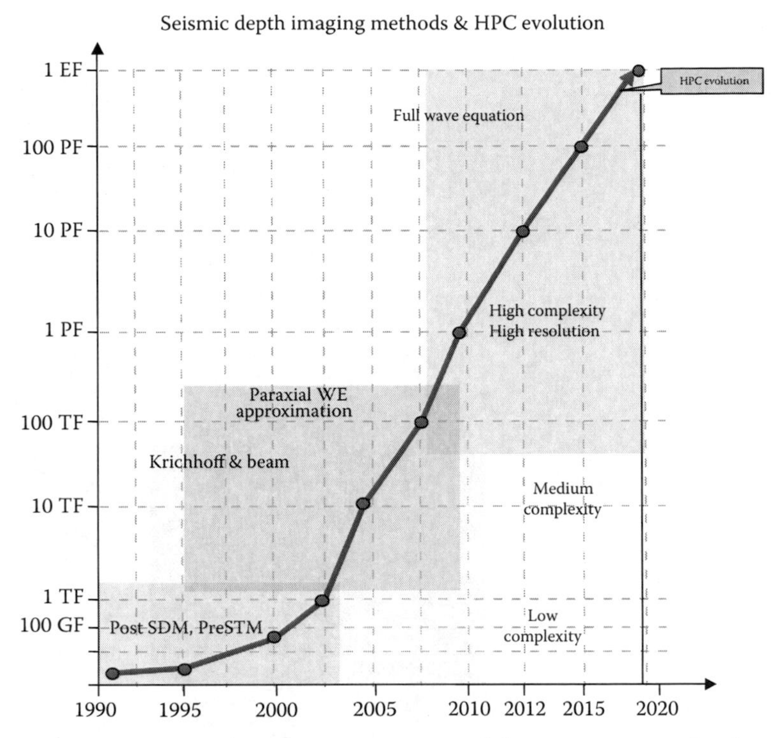

FIGURE 15.2 Estimated HPC power required for new seismic depth imaging methods (Calandra, Etgen, Morton). Note that each grid increase on the vertical axis represents a 10x increase in power.

return-on-investment justification, and modern improvements in digital software tools and processes need to be more effective.

ENDNOTES

1. http://en.wikipedia.org/wiki/Pennsylvania_oil_rush.
2. http://tle.geoscienceworld.org/content/24/Supplement/S46.full.pdf+html.
3. http://www.ieeeghn.org/wiki/index.php/Oral-History:Enders_Robinson.
4. Chapter 1: NCSA's Black Proposal.
5. An oil barrel (bbl) is 42 U.S. gallons or about 159 liters.
6. A spar is a type of floating oil platform typically used in very deep waters.
7. http://www.offshore-technology.com/projects/mad_dog/.org/wiki/Mad_Dog_oil_field.
8. For more information, see, http://www.bp.com/en/global/corporate/about-bp/bp-and-technology/more-discovery/land-seismic-imaging.html.
9. Full presentation, http://static.og-hpc.org/Rice2011/Workshop-Presentations/OG-HPC%20PDF-4-WEB/Henri-John-Scott.pdf.

CHAPTER 16

Rolls-Royce Corporation

High-Performance Computing Impact on Industrial Design Cycle

Todd Simons

CONTENTS

16.1 INTRODUCTION

High-performance computing (HPC) offers several key capabilities for industries involved in advanced manufacturing. Advanced design and simulation software provides engineers with the ability to analyze complex geometries and design components that meet challenging design requirements that could not otherwise be accomplished. HPC enables higher-fidelity analysis to improve and accelerate the design process providing improved products. Advanced analysis also supports the development of advanced products by reducing or eliminating physical tests through the use of virtual testing. This capability requires ongoing software development because of the ever-changing technologies in hardware and software used for large-scale parallel computing. These activities require support from national policies on HPC to advance science and technology and ultimately benefit society.

16.2 IMPACTING THE DESIGN CYCLE

The design process for components in high-performance applications is a multidisciplinary effort requiring expertise in multiple areas, including structural analysis, thermal analysis, aerodynamics, materials and process modeling, dynamics, lifing (part life analyses), manufacturing, and cost engineering. The design process must balance design objectives against requirements from these different disciplines, and these constraints are often at odds with each other. In aerospace, one of the primary design objectives is minimizing fuel consumption. Fuel costs are one of the largest operating expenses for airlines. Moreover, more fuel-efficient engines reduce greenhouse gas emissions from flight. One way to tackle fuel consumption is reducing weight. Minimizing material to reduce weight makes it more difficult to meet the requirements for satisfying strength and lifing. In a typical design cycle, a team of engineers takes a component through an iterative process where they study the trade-offs for different compromises that will ultimately meet all of the design requirements to produce an acceptable part. The numerical simulations used in detailed analysis can be time consuming and be a pacing item during the design process. Improvements in designs happen incrementally as the design team explores and analyzes the design space available. These design efforts cannot continue indefinitely because of project deadlines and cost constraints. Decreasing the time required for each design iteration allows the design team to more thoroughly explore the design space and improve the quality of the design. These time constraints make HPC and computing capacity a critical resource for design.

Gas turbine engine design begins with lower-fidelity models to explore the design space efficiently. Initial performance models start with one-dimensional (1D) models to represent the major components in the gas turbine engine. These models use empirical relationships, historical information, and simple models to approximate the behavior of potential designs. These models are used for evaluating engine subsystems such as compressors, combustors, and turbines, to size these components for materials used in fabrication, flow rates, pressures, air velocities, size, and number of airfoils. As the design process matures, higher-fidelity models are introduced. Two-dimensional (2D) models are introduced to evaluate aerodynamics on a plane at the mid-section of the gas turbine engine. These lower-order models are used to efficiently move the design in the right direction. Simplified 3D models are used in the early design process as well to validate that early design choices are viable and that a final design can meet the design targets.

In high-fidelity analysis, the equations that represent the underlying physics are used to study the performance characteristics of different designs and shapes. Decomposing the parts, or the air around the parts, into a computational grid allows us to solve these equations. Capturing more geometry features, increasing the number of points in the grid, and extending the computational domain can achieve even higher-fidelity and more accurate analysis, but this increases the amount of computational effort considerably. More detailed grids capture smaller design features that can have a significant impact on performance. As the design matures, more details are added to the models. These detailed models are larger and take longer to run, causing delays in the design cycle. Often the designers must balance the amount of detail in the models with the time required to complete these calculations. One of the benefits of using HPC is that the run times of detailed analysis can be reduced from weeks to days and, with enough computing power, from days to hours. The designers must evaluate gas turbine engines at different engine speeds and aircraft operating conditions, increasing the computing requirements. The more detailed the models and the longer the analysis takes to run, the less impact it has on the design. The analysis for design and simulation would ideally run overnight for the engineers and designers to review the following day. Overnight runs allow for a tight integration with the design cycle.

Figure 16.1 depicts the design maturity in a design cycle. The preliminary design process, which uses lower-fidelity and simple models early in the design process, creates a design with a lower maturity level. As higher-fidelity and more computationally intensive design tools are used, the maturity of the design increases. The increase in maturity comes as engineers develop more confidence in how the design will perform and as designers develop a better understanding of how component details impact the design. Design maturity also increases as the risks of not meeting design requirements are minimized.

Some problems in the design process can be simplified to allow for steady-state analysis, that is, analysis that for engineering purposes is relatively independent of variation in time. These are generally applied to single airfoil (wing or blade, such as propeller, rotor, or turbine) passages in the turbomachinery. Steady-state analysis can be applied to aerodynamic analysis, thermal analysis, and some stress analysis. The most

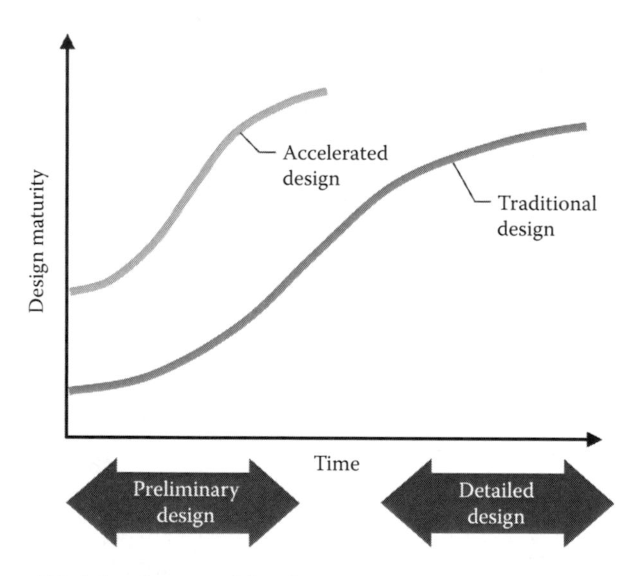

FIGURE 16.1 High level view of the design process.

computationally-intensive problems either evolve with time, or the physics of the problem require transient analysis. Both the gas turbine compressor and turbine subsystems comprise alternating rows of stationary and rotating airfoils that incrementally compress or expand the air. The relative motion of these rotating parts creates complex and unsteady aerodynamics in the gas turbine engine. Unsteady analysis of these components requires modeling multiple airfoils over both the stationary and rotating blade rows, which increases the computational size of the problem over an order of magnitude (power of 10). These unsteady calculations must be run until initial transients have passed through the domain and the solution reaches a cyclical steady point. Depending on the number of blade rows these calculations can require more than two engine revolutions for transient phenomena caused by imperfect initial conditions to pass through the domain. These problems often require days of analysis and as a result are not performed until very late in the design cycle. Figure 16.2 shows the airfoils in a multistage model used to set up unsteady calculations in a compressor. Flow moves from the left to the right. The colors indicate the pressure, where blue represents lower pressures and red represents higher pressures.

Modern three-shaft gas turbine engines have around 5000 airfoils. Numerical models for aerodynamics require around one million cells per airfoil to model the physics of airflow for performance analysis. Modeling techniques that capture the small holes used in film cooling require

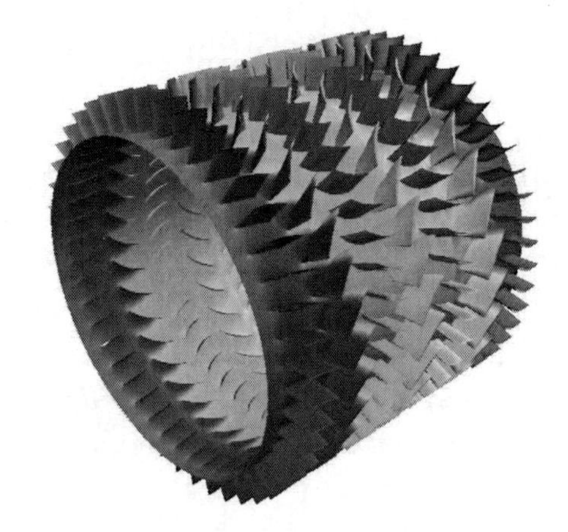

FIGURE 16.2 **(See Color Insert.)** Gas-turbine compressor with seven rows of airfoils.

significantly more computing resources than clean defeatured geometry. Multiphysics that include aerodynamic analysis along with heat transfer and film cooling in the hot section require many more grid points. Unsteady multistage analysis is needed to track hot streak migration from the combustor through the turbine section. Simplifying assumptions are used wherever possible in engineering to reduce the cost of analysis and to allow optimization of the dominant factors in the design. Numerical analysis of this kind of detailed geometry is simply not possible without HPC.

16.3 PHYSICAL TESTING AND RISK REDUCTION

The Federal Aviation Administration (FAA) and the European Aviation Safety Agency (EASA) mandate blade-off and bird strike testing for airworthiness certification before aircraft jet engines can fly passengers on planes. Blade-off testing is performed by placing an explosive charge at the root of a fan or turbine rotor and detonating the charge during test engine operation. During the test, the engine casing must contain the debris from the released blade to prevent damage to the aircraft fuselage. The engine must also stay attached to the wing and avoid vibrations that would damage the airframe. These are expensive tests of a destructive nature. These tests are placed near the end of an engine development program and failure on these tests is very costly, not only because of the amount of rework involved but also because of the schedule impact for the air-framer. Engine program schedules support test schedules for new aircraft and entry into service. New engine programs can require several years of development and redesigning major components has a large impact on schedules.

HPC is used with very large numerical models to simulate these certification tests. Large finite-element (structural) models are assembled that simulate the entire engine and model the bird strike and blade-off testing. These models are used to mitigate the risks of having large tests at the end of the development cycle, as they provide confidence to the program that the engine design will successfully pass certification testing. These simulations also provide insight and guidance into the design progresses. Figure 16.3 shows a mechanical model for whole-engine analysis of a fan blade-off event in a gas turbine engine.

Numerical simulations also reduce the cost of bringing products to market by replacing physical testing with virtual testing. With the improvement in accuracy of these numerical models some of the bird strike testing has been eliminated with the approval of chief engineers and certification authorities. Once numerical modeling has been validated to accurately simulate these extreme events, they provide a cost-effective alternative to some but not all tests in an engine program.

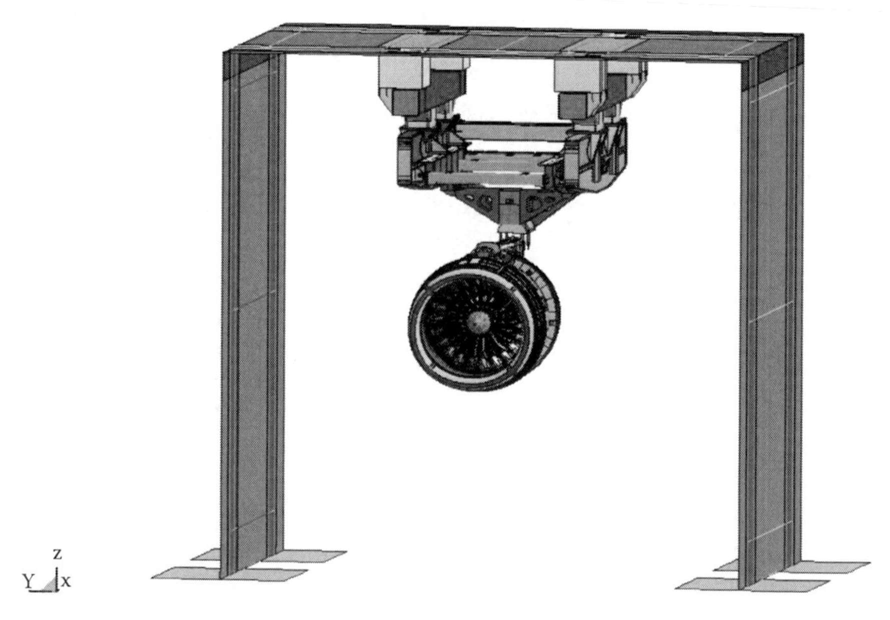

FIGURE 16.3 Large finite element model of a gas turbine engine.

16.4 SOFTWARE DEVELOPMENT FOR INDUSTRIAL PROBLEMS

HPC has been used for several decades in aerodynamic design and analysis by using computational fluid dynamics (CFD) and mechanical analysis using finite element methods (FEM). CFD, in particular, is suitable for HPC because these problems can be partitioned to run in parallel on multiple processors. Each processor can solve the equations of fluid mechanics and then exchange information with neighboring processors to synchronize the data between processors. CFD software programs continue to develop and are among some of the largest simulations performed on TOP500 supercomputers. CFD software can scale up to tens of thousands of processors. Implicit (time-dependent) FEM analysis of extreme event modeling, such as blade-off analysis, is more challenging to scale up to very large numbers of processors. FEM software programs have only recently scaled up to thousands of processors.

Many companies continue to develop their own proprietary software for complicated physics and for fluid flow that can be modeled with the Navier–Stokes physics equations. There are several motivations for maintaining a team of developers to support proprietary software. One motivation is to maintain technological leadership in an area of modeling and simulations. Another motivation is to customize the software for the specialized workflows required by the design process. By keeping the software development in-house, companies can better control software accuracy and variation between releases. Extensive testing done for verification (meeting design specifications) and validation (meeting operational requirements) is required for any software. The cost of internal software development must be weighed against the benefits.

Using commercially-developed software has the advantage of better user interfaces and better interaction with solid modeling software, but the licensing costs can be very expensive. As a result of large companies maintaining their own software, commercial analysis and simulation companies tend to market their products to small- and medium-sized manufacturers that do not have the resources to develop software internally and do not have extensive computing resources. As a result, there is a chicken and egg problem with commercial software not scaling to a large number of processors because their customers do not have access to a large number of processors. Here, the market place is dynamic. With the introduction of cloud computing and the ability to spin up thousands of processors for a short duration there will likely be a shift in the way commercial companies license parallel codes; this is already happening with some vendors.

Most companies have moved away from developing internal finite element codes. Commercial finite element software is adequate for most industrial problems of stress analysis, dynamics, and heat transfer. Companies take these products through validation sets before releasing them for production internally.

It needs to be stated that software development, the analysis of components, and HPC are a means to an end. The company is selling a product, and the analysis and simulation is a means to design a part that meets customer requirements that can be manufactured at a competitive cost. One of the objectives for any company is to minimize the cost of the analysis, the time to complete the analysis, and the cost of computing those results.

16.5 NATIONAL HPC POLICY AND INDUSTRY

Private–Public partnerships have been an important contributor to industrial and manufacturing development. High-technology manufacturing has benefitted from investments in national supercomputing centers. Investments in leadership-class machines provide platforms that allow researchers to push the scale of both industrial and commercial codes.

HPC is a specialized skill that is in growing demand. Finding qualified specialists in HPC is a challenge for industry. National computing centers provide an environment where specialists can be trained, but also where industrial domain scientists can climb the learning curve for adapting software to run efficiently in large computing environments. Collaboration between HPC specialists and domain scientists is needed to adapt codes to run efficiently on evolving hardware platforms. The move to many-core processors and the adoption of accelerators, such as graphics cards and coprocessors, requires some restructuring of the software to achieve the best performance. The software tools and libraries used to scale parallel software up to new hybrid computing models are constantly evolving and require ongoing training for software developers. Companies that have made a considerable investment in validation want to be able to extend their codes to new platforms without changing the results generated between different versions and platforms.

The other benefit of collaborating with national supercomputing centers is that industry can evaluate different hardware and get independent expertise by testing proprietary software on new hardware platforms.

Investment in HPC is challenging in many industrial organizations, and these computing centers provide industry with the computing experience to make informed decisions about major hardware purchases.

Staff scientists and researchers at national facilities also benefit from the interaction with industry to work with current models and problems from the field. This collaboration meets the objectives of supporting the industrial base and contributing to science, to the economy and to society at large. By helping companies advance modeling and simulation, better products are brought to market. Advances in gas turbine engines improve the safety and reliability of travel, improve the ability to move products globally, and improvements in fuel economy reduce the production of greenhouse gases.

16.6 CONCLUSION

HPC has become an important enabler for the design and analysis of high-performance components and for gas turbine engines. One benefit of this technology is allowing high-fidelity modeling to increase the accuracy of design calculations in a time scale that can impact the design cycle. Parallel computing is required to scale up the simulation of large models to allow solutions to be generated in a timely manner. By placing high-fidelity modeling earlier in the design process a higher degree of maturity is reached earlier, design schedules are reduced, and design costs can be minimized. Design improvements deliver more fuel-efficient engines that benefit society by reducing greenhouse gases.

Another benefit of HPC is risk reduction associated with certification testing in gas turbine engines. Virtual testing of bird strike and blade-off tests provides confidence to the engine programs that the designs being developed will be successful. In some cases, virtual testing can eliminate instrumentation or reduce the number of tests required.

Efficient use of HPC requires ongoing development of software and collaboration with software vendors to support modern computing hardware. Computing hardware continues to evolve. National policies supporting HPC are needed to provide support for domain scientists to advance software and hardware development. National computing centers provide a valuable resource for industry where collaboration can take place between industry, academia, software vendors, and hardware vendors. National computing centers also provide training and support for modifying software to run efficiently on modern computing architectures.

CHAPTER 17

Porsche: An Industrial Use Case

Michael M. Resch and Andreas Wierse

CONTENTS

17.1 THE ROLE OF HIGH-PERFORMANCE COMPUTING FOR THE GERMAN AUTOMOTIVE INDUSTRY

The German automotive industry is operating in a special environment. Salaries in Germany are relatively high compared to international competitors and especially compared to emerging economies. Very early it was clear that international competition would put the manufacturing of cars in Germany at the risk of being outcompeted by low-wage countries. In this situation, the path chosen was one of clear focus on high technology and high quality. Simulation was a cornerstone of this strategy right from the beginning. Starting with the Cray systems that were installed in the 1980s, German car manufacturers started to grow their high-performance capacities. Today, HPC is an integral part of their IT strategy with a trend toward extending the reach of simulation, both horizontally and vertically. Originally utilized for crash simulations, HPC plays a major role in all stages of the development today. Furthermore, HPC has branched out from development departments into production planning and, in some places, even production control.

17.2 PORSCHE AS A SPECIAL CASE

For Porsche—as the smallest of all German car manufacturers—high-performance computing (HPC) was especially important but also a huge investment. Given the need to maintain a steady pace of development and the relatively small number of units sold, simulation was very early on considered to be a chance to survive in an increasingly competitive market. So, for Porsche, simulation was one of the cornerstones in overcoming the disadvantage of being a small and independent vendor.

As Porsche traditionally already had a very good relationship with the University of Stuttgart, Porsche started to collaborate with the computing center of the university in the second half of the 1980s. Many of the key engineers at Porsche were alumni of the University of Stuttgart and the Porsche development center was close to Stuttgart. Common development in engineering between Porsche and several institutes of the university had a long tradition. Proximity and the trust that had been built over decades of collaboration provided an excellent breeding ground.

First, the main focus was on crash simulation and common projects. The ever-growing need for computing power soon led to the first cooperation. The University of Stuttgart was hosting Porsche systems as early as 1993. They were still owned by Porsche but were integrated into the operational environment of the computing center. As much as this was an early success, it soon turned out that a more stable model for cooperation had to be found, as with every new acquisition of a system new arrangements had to be found to guarantee industrial access.

17.3 SUSTAINABLE HPC PROVISIONING THROUGH A PUBLIC–PRIVATE PARTNERSHIP

With the foundation of Hww* in 1995 as described in Chapter 5, a platform was provided to continuously make research systems available to industry. At the same time the High-Performance Computing Center Stuttgart (HLRS) was founded, which served as the entry point of Porsche into academic HPC. Through HLRS and Hww, Porsche had access to world-class resources with a focus still mainly on crash simulations. Issues of networking and security had to be resolved but solutions were found based on mutual understanding and exploiting improved network and security technologies as much as possible.

* Hww stands for Höchstleistungsrechner für Wissenschaft und Wirtschaft GmbH, which is the interface between HLRS and industry.

In 2005, Porsche decided to extend its product line with the creation of a fourth platform called Panamera. Right from the beginning, the decision was made to use HPC resources of HLRS/Hww as much as possible. Still the main issue was crash simulations, but computational fluid dynamics and acoustics gradually gained importance (Figure 17.1).

The Panamera project became a touchstone for the collaboration between Porsche and HLRS. In a very short period, a number of issues had to be tackled to make sure that HPC simulation could easily be integrated into the development processes of a successful and demanding car manufacturer. Things were made easier as people on both sides knew each other for more than 10 years already. Flexibility of HLRS staff on the one hand and the organizational stability of Hww on the other hand helped to create a working environment that allowed Porsche to rely to a large extent on the computing services of HLRS within a few months from the start of the Panamera project. As a consequence, industrial usage of HLRS systems grew dramatically and reached 100 million core hours by 2013, which resulted in an annual turnover of about €2 million—the biggest share coming from Porsche.

With the success of the Panamera project, it became clear that both sides wanted to extend the collaboration. Pure computing time was not enough. Porsche realized that it could benefit from the know-how of

FIGURE 17.1 Porche Panamera.

HLRS as well as some of the users of HLRS. For HLRS the collaboration required a further deepening through an extension of services into the field of applied simulation. Discussions about a way to bring research and automotive industry closer together were initiated by Porsche and HLRS in 2006. Other German manufacturers were approached and it became clear that software and hardware vendors should join. As an offspring of their long-term relationship, the Automotive Simulation Center Stuttgart (ASCS) brought together three German car manufacturers, more than 10 independent software vendors, 3 hardware vendors, and several research organizations in 2008.

17.4 SUMMARY

The use of HPC by Porsche, based on intensive collaboration with HLRS beginning in 1993, is based on the unique combination of a number of factors. Proximity played a vital role. It also provided political support as state politicians were interested in keeping Porsche connected to Stuttgart as much as possible. Trust was another key factor. The fact that many engineers working at Porsche were educated at the University of Stuttgart was extremely helpful for both sides. Trust was also built up during the Panamera project. The flexible handling of Porsche requirements by HLRS staff helped to overcome roadblocks and any mental reservations that still might have existed. The collaboration also triggered further developments like ASCS. Keeping up this excellent work is a mission that drives both IT experts at Porsche and HLRS. Trust is the capital on which to build the next steps.

CHAPTER 18

Industrial Applications' Journey to Supercomputing at Renault

Yves Tourbier and Marc Pariente

CONTENTS

18.1 INTRODUCTION

France's Renault Group has been an automotive industry pioneer in the usage of high-performance computing (HPC) since the 1980s. A yearly process is utilized to measure the needs and schedule the evolution of internal HPC capacity. Optimization studies, especially those on the upstream projects, are used to size the breaks on HPC capacity because they are representative of anticipated needs of projects in 5 years.

Concerning the structure of the vehicle body, innovations are systematically designed and tested with the help of numerical simulation augmented by validation with a physical prototype. That gives a lot of optimization studies with data-heavy numerical models such as crash, acoustics, and durability—a great number of design parameters (>100 in our study)—many real-world industrial constraints like assembly process or carryover maximization, and very complex objectives because all digital design problems requires compromise between dozens of contradictory specifications. This case study is a review of car body optimization and the trade-offs associated with mass and specifications with combinatorial constraints.

Renault has no intention of directly developing simulation tools, using commercial tools and implementing collaborative projects with software vendors and French laboratories to improve the performance of these tools. Renault uses national or European (through PRACE) HPC facilities in collaborative research projects, while numerical simulations for the vehicle or powertrain projects use in-house HPC only. As a result of these collaborations, Renault funds new PhDs every year in the field of numerical simulation. Renault also participates in an IRT (Institut de Recherche Technologique) SYSTEM X project dealing with the use of Model Reduction and Multiphysics Optimization.

18.2 CASE STUDY

18.2.1 Project Background

This is a world first-ever in this domain: neither Renault, simulation vendor ESI, nor their competitors, had ever used a head-up crash numerical simulation of this size in a combinatorial optimization study with such levels of parameters.

18.2.2 Critical Role of the HPC Community

Renault has been one of the first automotive companies to use HPC for crash or CFD simulations; this allowed Renault to stay in the forefront of the European EuroNCAP rules for safety certifications. By developing more and more realistic crash models, including fine-resolution description of all the parts of a car and detailed crash test dummy modeling, Renault has been able to embed a wide range of software for developing appropriate models and tools to reduce the number of physical tests and, by consequence, the cost of development and the overall time to market. This role of Renault and other car suppliers has been mandatory for developing a strong expertise in some of the leading automotive simulation vendors like ESI Group in France.

18.2.3 Business Opportunity When Using HPC

Renault uses HPC to reduce the delay of simulation optimization studies, affecting both time-to-market and improvements in car safety required by 2015 EuroNCAP6 regulations. HPC is also an opportunity to introduce more physics into models while keeping a good elapsed time in the design process (which has a very strict scheduling).

The study uses a crash simulation of a vehicle against a standard aluminum barrier. Optimization has been completed using the Design Of Experiments (DOE) method. The main difficulties lie in the dimension of the optimization problem (around 100 parameters) comparing to the cost of the crash simulation (24 hours with 1024 cores for each individual simulation), and the carryover constraint. The objective underlying the carryover constraint is to keep unchanged a given percentage of the parts from the old vehicle into the new one and to reduce the engineering cost and investment. The thickness of a part in carryover set can vary slightly; a part not in carryover set can change dramatically.

This very advanced project was not possible using internal Renault facilities, so Renault, ESI Group, and Ecole des Mines de Saint Etienne jointly applied to PRACE Open R&D calls for proposals and were granted successfully with more than 42 million core hours on the GENCI Tier-0 system called CURIE. Up to now, this capacity and capability allocation is the biggest allocation for an industrial user since the inception of the Open R&D program by PRACE in 2012.

18.2.4 Data Processing and Visualization

Implementation of a complex workflow between Renault and PRACE facilities:

- Preprocessing within Renault HPC facilities for coupling stamping process and crash simulations
- Computations performed on CURIE system
- Some post-processing on the PRACE HPC
- Other post-processing on our laptop
- Data visualization with Renault and PRACE facilities (remote visualization)

18.2.5 Advanced Manufacturing, Big Data, and the Industrial Internet

In our study, each crash simulation provided 24 GB of data with our post-processing system extracting the most important information needed to verify the specifications. The data generated will be used in data mining studies as part of the IRT System X ROM project.

18.2.6 Lessons Learned

1. ESI has strongly improved its PAM-CRASH scalability.
2. Renault has shown that HPC can handle a problem with a large number of decision variables (>100 Boolean variables: 1 part is carryover or can be modified) on heavy calculations within a reasonable amount of time.
3. The optimization can effectively call the maximum number of calculations in parallel.

18.3 WHAT'S NEXT AND CONCLUSION

This first large-scale study made with PRACE facilities will be now reproduced into Renault engineering process (using Renault HPC facilities and with a smaller crash simulation) in the next few months, and standardized in future vehicle projects (Figures 18.1 and 18.2).

Work will continue on crash model reduction to further reduce the cost and delay of this kind of optimization study. An example of such results is as follows:

The study lasted several hundreds of head-up crash simulations: 24h and 1024 cores per simulation on CURIE.

FIGURE 18.1 A virtual Renault.

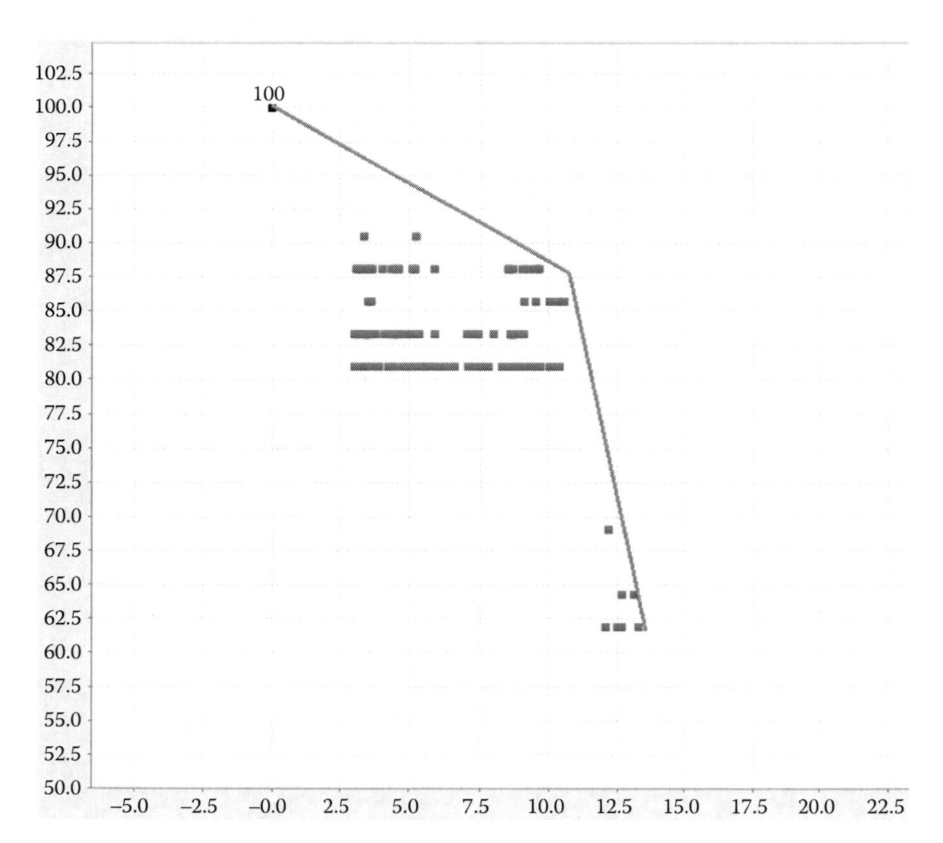

FIGURE 18.2 The black dot named "100" is the starting point of the optimization. The reference vehicle at "100" represents zero change in mass. Y axis indicates % reuse; X axis represents weight gain in kilograms.

Optimization of the reduction in weight of a new car (x axis) is a function of the percentage of component reuse, or the pieces of the car that are reused (y axis). Higher reuse from previous generations lowers costs since new parts do not have to be designed. However, better performance is obtained by reducing overall weight since this also reduces CO_2 emissions.

Each middle dot corresponds to a single crash model simulation performed on CURIE, and only the valuable solutions are shown here. Between the initial point (on the top left) with 100% reuse and no gain in weight and the last point (on the bottom right) with more than 12.5 kg of weight gain but with only 60% of previous parts reused (which is too expensive) it is possible to draw a Pareto front for guiding the optimization process. The knee in the Pareto represents the point at which additional weight gain from crash-worthy new parts becomes increasingly costly. A 10 kg gain in weight is optimal here with 85% of the parts unchanged.

CHAPTER 19

GE Research Industrial Applications' Journey to Supercomputing

Rick Arthur and Masako Yamada

CONTENTS

19.1 GENERAL ELECTRIC'S TRADITION FOR INNOVATION

World-wide, the name Thomas Alva Edison is synonymous with *Invention*. In founding the General Electric Company (hereafter "GE") in the 1800s, Edison created an institution for industrial innovation. Over the past century, engineers and scientists at GE have shaped numerous technological transformations in our lives. Starting with technology to generate and transmit electricity to businesses and the home, that electricity then powers GE lighting, motors, and appliances. GE pioneered the technologies enabling the broadcast of radio and television to the home. GE jet engine advances endowed military and civilian aviation with speed, range, and efficiency. GE technology leadership in diagnostic imaging (computed tomography [CT] and magnetic resonance imaging [MRI] scans) has nearly erased the phrase "exploratory surgery" from the practice of medicine.

Now, 120 odd years later, GE operates in over 100 countries with a strong and persistent presence spanning not only the heritage businesses but now also complex industries like locomotives, water treatment, financial services, life sciences, and healthcare IT. This breadth of business areas, each distinguished by a depth of technology, creates the microcosm in which GE Global Research[1] practices science and engineering. GE Global Research serves as the company's hub for research & development, works closely with the technologists in the businesses to innovate improvements and entirely new products, often leveraging experiences from one industry to another (for example, using medical CT scanner technology for inspecting parts for a jet engine). Likewise, essentially all products and services apply numerical tools or integrate computing components, affording opportunities for common investments to deliver diversified impacts.

Another term synonymous, in fact eponymous, with GE's founder is *Edisonian*,[2] based on a perhaps misleading record of the method of invention used in discoveries of materials needed in electrical lighting, microphones, etc. However, the culture associated with systematic search through empirical trials has indeed been a hallmark of GE's research and development. There are two very sound reasons for this: many of the relevant sciences lacked robust comprehensive numerical models to replace physical study, and second, many of GE's products are by their nature critical to public infrastructure and/or human health and safety. As such, the burden to justify costs and time to use numerical methodologies presented challenges due to incompleteness and uncertainty inherent in limited-fidelity modeling. Moreover, regulatory agencies governing clearance for product release offered few to no incentives to use it either. In several areas, limitations to modeling capabilities or physical understanding still compel empirical methodologies; quantum effects and biology being prime examples.

19.2 GE AND COMPUTING

The nature of GE's highly technological products demands computational access—whether built internally or purchased from others. Some may recall that GE was a mainframe computer manufacturer in the 1960s with its Compatibles/200 line[3] until selling that business to Honeywell in 1973. Today, only GE Intelligent Platforms sells a computing-only product: ruggedized platforms (including high-performance computing [HPC]) for harsh environments. Computers are integrated into many GE products such as embedded controllers (operation of jet engines), for data acquisition and processing (remote health monitoring of power generators), and to enable end users to interact with processed data (medical images to radiologists). Although these use state-of-the-art technology, typical designs fall into the gigascale to low terascale computational category.

Advancing computing capability to meet business needs along a winding road of acquisitions and divestitures over the years, GE researchers achieved world-class computational recognition from work in operating systems,[4] computational complexity,[5] graph theory,[6] computer graphics,[7] and object-oriented modeling.[8] The impact has reached far beyond GE or the government contracts on which GE developed them to broadly influence cutting-edge artificial intelligence, industrial controls, design automation, signal processing, and visualization. Such expertise provided groundwork for experimentation with modeling and simulation in product and system design leveraging high-end vector machines and mainframes through the early 1990s.[9]

19.2.1 Shift to Modern State of the Art

The introduction of highly capable workstations[10] and clustering technology in the 1990s enabled wider adoption of modeling and simulation in industrial product and system design. The computational power, affordability, and ease of use radically changed the body of potential users, in particular by lowering the investment and risk associated with the effort relative to the value of speculated results. This computational power, however, only weakly justified a reduction of capital and labor costs; as modeling and simulation data still lacked the fidelity and scale to obviate physical empirical experiments.

In the design and manufacture of advanced technologies such as medical imaging, aircraft engines, and power generation and distribution, GE must be uncompromising in safety and reliability. Confidence in the data from numerical modeling is paramount to its utility and uncertainty forfeits potential product advantage. The engineering team must compensate for the unknown, for example, lowering operational temperature or over-engineering the design by adding alloy thickness in a part. The former reduces competitiveness and the latter impacts time to market, cost, and other performance metrics of the system (such as fuel consumption due to additional weight).

In pursuit of more competitive designs and a more agile process, engineers persistently need not only hardware with significantly greater computational horsepower but also software tuned to answer the specific questions faced by design engineers. These questions range from highly iterative detailed final designs to massive scale reference designs spanning multiple subsystems and often multiple physical models. To help address these questions, GE initiated an Advanced Design Tools (ADT) program to leverage commodity cluster HPC through internally developed solvers as well as partnerships with third-party software vendors. One essential activity performed by the program was continuously validating model data against results from physical experiments toward understanding uncertainty bounds and tuning numerical assumptions in the models. Physical tests are typically conducted for traditional engineering learning and as required for regulatory certification. The data collected from these tests are highly valuable in qualifying and quantifying the numerical models as well. The tests are carried out at numerous locations throughout GE. Many require highly specialized skills, facilities, and in some cases tremendous costs (equipment purchase or custom build, instrumentation, setup costs, data analysis costs, and some tests are destructive in nature.)

Once the software and data for the models were carefully validated for accuracy, these became an integral part of the design engineering process from subcomponents through system integration. However, ambitious users requesting large proportions of the shared computing infrastructure would find that there was a practical maximum limit to requests, well below the overall system size. To ensure their job submission would actually be scheduled and completed within needed design iteration timeframes, users would adapt their models and simulations to adequate but reduced size. The ramifications of these computing constraints included strategies like separation of physical models, separation of geometries, separation of systems, and simplification of detail. GE benefitted from the computing industry's dramatic improvements over the past decade in both aggregate computational capability and per-operation affordability at scale that has unfettered such constraints and led to reevaluating the viability of multiphysics, multisystem, and highly detailed geometric modeling. The success of the ADT program in leveraging these advances in computational hardware and modeling software resulted in the formalization and growth of a team dedicated to custodianship of the tools and infrastructure. This team carries the name GE Engineering Tools Center of Excellence (ETCoE) and manages the engineering software, hardware, and processes from workstations to large computing clusters.

19.2.2 Adoption of High-End Computing: Barriers and Successes

In 2005, the leader of GE's Computing & Decision Sciences technology domain formed the Advanced Computing Lab (ACL) to focus on adoption of emerging computing platforms across GE's businesses and partners and to assess technical disruptions. The group included a number of researchers engaged for many years with GE Aerospace platforms, which, through mergers and acquisitions, became Martin-Marietta and ultimately Lockheed Martin.[11] In order to maintain world-class competitiveness, ACL leveraged its familiarity with cutting edge computing and software employed in military/aerospace applications and its deep relationships with technology vendors to initiate a transition from commodity to more advanced computing platforms across a broad range of traditional applications. This lab's leadership offers the following perspective on the road to adoption of supercomputing, including a highly publicized case study written by its principal investigator, Dr. Masako Yamada, one of the lab's senior scientists.

To better understand historic, current, and potential use of numerical modeling and simulation at GE Global Research, a broad canvas of its technical thought leaders were surveyed. Patterns emerged from these conversations to form a list of categorical barriers to adoption. We separate these barriers into "technical" and "nontechnical" categories as shown below in Table 19.1. Technical barriers involve problem solving, may require invention, and thus have inherent uncertainty. The nontechnical barriers' challenges are more political and economic. Although these may be no more tractable than the technical barriers, plausible solutions could be reached through willingness to come to business agreements or by justifying the reasoning or courage to make an investment of time and/or money. We will describe the identified challenges, anticipating these are far from unique to GE.

Validation is key to avoid the so-called "Garbage-in/Garbage-out" trap. Addressing technical barriers generally falls to the community of subject-matter expert scientists and engineers working in concert with computational scientists. For example, to stimulate acceptance of numerical models, a robust validation and confidence quantification methodology is needed to harmonize simulation data with those resulting from empirical experiments and to quantify confidence bounds. Thus a process by which real-world tests can improve model fidelity and quantify model uncertainty must be developed and employed, and then a statistically robust methodology for uncertainty carried through the numerical modeling. GE's ETCoE has the facilities, scale, and reach into a breadth of technical talent for such validations and measurement. Although scarce, certainly government laboratories and several other industrial institutions beyond GE carry out similar critical validations. The persistent need for achieving adequate uncertainty quantification to enable engineering decision making or scientific conclusions often introduces additional complexity into models.

Complexity will often manifest in high-fidelity models, particularly those using first principles,[12] resulting in models that are computable in theory but not in practice. Some fields routinely use first principles (e.g., optics) because these are well understood and the computations are tenable on available systems. Others may be able to express a detailed

TABLE 19.1 Survey Results: Barriers to Adoption of Simulation-Based Engineering & Science

Technical Barriers	Non-Technical Barriers
Model and data	Resource availability
Validation	Software licenses
Complexity	Computing hardware
Solver/software	Cultural mindset
Availability & scalability	Legacy "Lock-in"

model as such (e.g., combustion) but no system in the foreseeable future could process the model in any relevant timeframe for incorporation into design practices. Still other systems are so complex as to be impossible even to describe numerically without resorting to stochastic abstractions and numerous approximations and assumptions (e.g., biology).

Availability presumes not only a validated and computable model, but also the existence of a software implementation of that model for the practitioner capable of reading and writing data at requisite detail and fidelity. *Scalability*, fidelity, complexity, or validation for adoption may still present significant barriers to adoption of modeling in some technical fields. The technical barriers stemming from solvers typically involve inherent limitations to the implementations such as the maximum size of a system the solver can model (e.g., 100 billion cells in a mesh), the solver's lack of parallel efficiency at desired scale (e.g., 100,000 concurrent threads), or the ability of any solver to simultaneously model the desired multiple physical models or scales (e.g., fluid mechanics plus combustion chemistry). This is an artifact of many codes having been conceived long before modern architectures (e.g., massively parallel, heterogeneous) were imagined as platforms of choice for their users. It would have been quite rational in the early 1990s, for example, to assume enormous clock speed improvements throughout the useful life of a particular software implementation.

Where such gaps have become evident, software communities comprising users and developers (often intermixed) rely on persistent dedication of tremendously innovative, motivated, and skilled professionals. Some commercial software products provide exceptional modeling capabilities, though their scalability is highly dependent on their target market. High-end users such as GE often are a tiny niche of a software vendor's customer base. Where profit motive has not aligned commercial vendors to provide sufficiently scalable or capable products, GE has turned to internally developed, open-source or public domain software. Internally developed software typically focuses on the specialized nature of the problem to GE's products, thus investments and feature prioritization are directly driven by the value of competitive technical advantage. GE often uses more general open-source and public domain software in collaborations or when executing government contracts. The usefulness of these noncommercial codes varies widely to GE. For example, while some of the public software, such as that resulting from the Department of Energy's (DOE's) Advanced Simulation and Computing (ASC) program,[13] is capable of unrivaled scalability and generates experimentally validated results, some university-based codes with the most ambitious capabilities cannot be considered for industrial use because of weak or absent formal validation with data derived from physical tests.

Software Licensing Cost was the most common nontechnical barrier expressed by technical practitioners. Many of the commercial engineering software vendors use cost structures based more on 1990s-style use, tying licenses to seats and processors that poorly map to modern problems at (multimode + multicore) scales of interest. Consider the difference in a $500-per-core license being run by a single user on their 8-core desktop system vs. a 1000-core cluster (while costing the software vendor no more). Pricing schemes targeting production users presume that buyers amortize solver costs over the engineered product or service. Yet, researchers aimed at proofs of concept, ironically for production users in most cases, would be unable to experiment at the edge of scale and capability merely based on the prohibitive price. Often exploratory projects at GE Global Research are conducted on tiny budgets only enabling jugular experimentation, and even ongoing projects may be more opportunistic than strongly tied to a product line. Such financial structuring will strongly bias would-be users away from attempting modeling unless alternative options exist, such as open-source software or vendors offering trial or unlimited use models. This became much more prominent where the minimum numbers of cores to run a relevant model for a problem numbered in the tens of thousands.

Computing Hardware limitations as a barrier to adoption were not always directly observable. The fairly accurate assumption that any hardware expansion will be immediately consumed without diminishing the outcry for more by insatiable users created a culture of compromise that doubled as background noise in hiding some missed opportunities. A great sum of important work of scientific and engineering interest would fit on a high-end workstation or the internal ×86 cluster. An exceptionally capable HPC IT support team maintained state-of-the-art, cost–effective, robust computing infrastructure and carefully monitored a number of metrics used in year-over-year planning. But as job requests began to rise into several hundreds of cores, the wait times for execution extended beyond intervals for which the computed results would be relevant (e.g., a design iteration). Of necessity to complete their work, users would adapt to the system limitations by trimming the sizes of their jobs to something they believed would be more likely to be successfully scheduled, or targeted external platforms (which will be discussed in Section 19.2.4.). There were no surge demand options, and because of the self-limiting users, for some time the

IT-collected metrics did not bear out this missing potential demand, and thus the lost opportunities. Commercial on-demand computing options have only recently had the requisite system capabilities for such simulations, and even if these options were capable, the nature of most of the targeted work would be too proprietary to accept the data protection assurances of these entities (from a legal and/or IT security standpoint.)

Cultural Mindset held long-established and deeply engrained physical, empirical science bias within some organizations. Although the ADT-driven turbomachinery businesses (engines and power generation) had evolved a cultural acceptance for numerical modeling, nearly all other fields at GE still held strong Edisonian empirical experimentation preferences, for the reasons cited earlier. To address the very valid criticisms in such fields, the question of scale became centerpiece to the discussion—as noted external successes[14] showcased that the inability of numerical simulations to accurately model phenomena could be overcome once a critical threshold of scale allowed the scope of the solver and data to encompass the needed physical system. While there remain physical systems difficult and perhaps impossible to characterize, a discussion of acceptable scale would evolve instead into a discussion of hardware and simulation software cost.

Legacy "Lock-in" ironically impacted the otherwise progressive users of the ADT program. The successfully used simulation practice created a significant legacy of modeling tools, practices, and regression data that had become a barrier to adopting newer algorithms, computing platforms, and scaling capabilities—particularly in regulated industries where recertification of the tools and processes represented a nontrivial effort or expense. But having accepted and proven out the value of numerical modeling, these talented engineers and scientists were also the most ready for the leap to supercomputing and high-fidelity models.

19.2.3 The First "Ah-Ha!" Successes: Insights into Portability and Scalability

The perceived costs not only of HPC platforms, but also in software porting (internal codes) and licenses (vendor codes) delayed GE's adoption of terascale modeling. Each ADT program must rigorously justify its costs based on impact to new product launch or deployed fleet products. The codes run at this stage were typically tens or hundreds of cores over hours to weeks for results. To reach the goal of expanding from individual components to subsystems toward linked multiple subsystem modeling would require entirely new computing strategies. In evaluating migration to platforms such as IBM Blue Gene (BG)/L or Cray XT5, several stakeholders asserted concerns over unknown costs and complexities in adoption with unknown performance benefit. Similarly, a number of known problems of interest were not even considered for numerical modeling because of a minimum threshold of complexity necessary for relevant results.

The initial strategy toward disruptive impact was to identify forward-thinking individuals on high-visibility programs and enable their success at a scale beyond the presumed and accepted limitations. My companions in this part of the journey shared my pure stubborn persistence and a patient certainty in the cause. Emboldened by successes[15] at nearby Rensselaer Polytechnic Institute (RPI) a skunk works project ported a key GE internal code to an IBM BG/L in only a few days. Although it ran correctly, it had not been tuned to run efficiently. A fortuitous meeting between GE and IBM executives resulted in an opportunity for IBM's highly talented HPC team to evaluate that code and leverage BG/P machines to which IBM had internal access to demonstrate the potential scalability. This effort's success enabled GE to explore relationships with the massively parallel systems of the Department of Energy's Office of Science.

Leveraging a wide range of internal cross-organization talents and external connections to successful practitioners, a few early success demonstrations won over key executive leaders at GE Global Research, including the chief financial officer, the Advanced Technologies vice president, and the technology director for Aero-Thermal and Mechanical Systems. Once the leadership team shared the passion to champion a dramatic change in the scale by which numerical modeling would be used in our research methodology, the years spent in the dark forest of skeptics were redeemed.

In 2009, GE engaged a number of supercomputing facilities to perform a study to benchmark HPC system scalability. The benchmark was significant at the time in that it represented the operations of an actual industrial code rather than using a traditional scaling benchmark like LINPACK. The results were shared with all of the participants, though never formally publicized. The most direct goal was to test and quantify anecdotes claiming a practical limitation on scalability of commodity clusters as compared to specialty architectures such as IBM BG and Cray XT. The test was devised by GE and the agreements were made with supercomputing facilities to provide computing time for performance measurements. The participants and platforms included IBM Research (Yorktown Heights BG/P), RPI (Computational Center for Nanotechnology Innovations BG/L), NCSA (×86 Intel cluster), and GE Global Research (internal ×86 AMD cluster).

The results confirmed that the IBM platform's engineered reliability and superior network fabric provided far better weak and strong scaling[16] despite a significantly slower clock speed. More importantly, the data illuminated (2009 era) ×86 cluster reliability effectively and limited aggregate problem size to roughly 2000 processor cores. As it was clear, GE would need to run well in excess of this capacity, the insight into the reliability limitation in itself justified the need to invest in the leap to high-reliability systems.

19.2.4 Engaging the U.S. Government's HPC Ecosystem

The products of the IBM collaboration allowed GE to leverage the BG/P "Intrepid" deployed at Argonne National Laboratory to measure scaling efficiency beyond what was available at IBM's internal facilities, as well as to improve the results of a modeling effort for which GE was contracted by the U.S. government. The scaling study then also expanded to include the Cray XT5 "Jaguar" at Oak Ridge National Laboratory, completing the target landscape of computing platform alternatives. The results of the benchmarking were once again shared with the study participants.

GE's advocates for supercomputing-scale modeling now needed a showcase problem to highlight a significant leap in engineering value enabled by such platforms. To justify internal investments in hardware and software for high terascale, a scalable solver would need to demonstrate a previously unachievable result on a relevant engineering problem. To meet the government facilities' computing grant requirements, that problem would need to use nonproprietary data and its result be of general scientific interest and publishable. To ensure legal readiness to use the government facilities, extensive effort was applied in getting user agreements in place between GE and the Office of Science laboratories. This was devised not only as groundwork for accepting grants of nonproprietary computing time, but also proprietary use in the event of an acute business need. In 2010, six potential "breakthrough" problems were identified and explored toward candidacy. One problem won GE's first INCITE award,[17] and two won Director's Grants.

Under one of the Director's Grants, an important jugular test was conducted on the Cray XT5 Jaguar at Oak Ridge National Laboratory (ORNL). The simulation modeled fluid dynamics of the low-pressure turbine (LPT) component of an aircraft engine design. Not only was the LPT simulation performed at a geometric scale previously unthinkable but it included unsteady calculations never before modeled (Figure 19.1).

The results of this test are shown in Figure 19.2. The colors correspond to entropy in fluid flows, whereas the white slices represent cross sections of the metal "blades" of the turbine, around which these flows were modeled (from inlet on the left to the exhaust on the right). Although single engine blades had been modeled for decades, expanding the scale of the model geometry to incorporate blade row passages and in particular multistages of such passages already pushed this simulation well beyond the practical capabilities of GE's internal HPC facilities. If such a model were to be run at GE, the result would have looked like Figure 19.2a as a time-averaged steady-state representation. You should first note an improvement in data fidelity when comparing that data as in Figure 19.2a with the previously unattainable model complexity of the dynamic unsteady analysis shown in Figure 19.2b. The addition of the strut wake to the model in Figure 19.2c, represented within the blue inlet flow by the green-to-red band, brought to light previously unknown engineering insight regarding its persistence through the LPT. Expanding the model scale and complexity to a crucial threshold (180,000 iterations over 200,000,000 grid cells) enabled a scientific observation only achievable through large-scale simulation, and this jugular experiment achieved the objective of conclusively catching the needed attention and support of GE executive sponsors. This project was awarded the IDC HPC Innovation Excellence Award in 2012.

This work also heightened awareness of the value of partnering with the DOE labs and led to a number of follow-on grants. These grants being nonproprietary relied on having access to NASA data sets or other

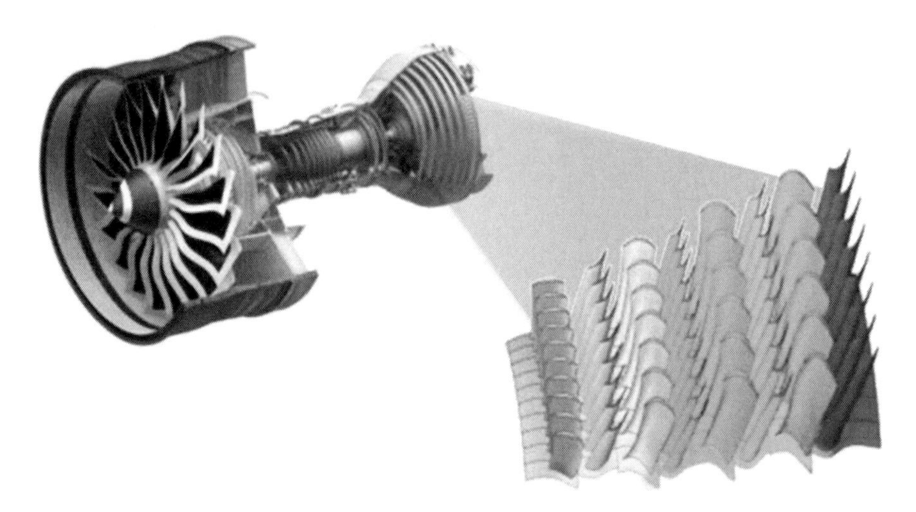

FIGURE 19.1 **(See color insert.)** Model of "blades" in the low-pressure turbine.

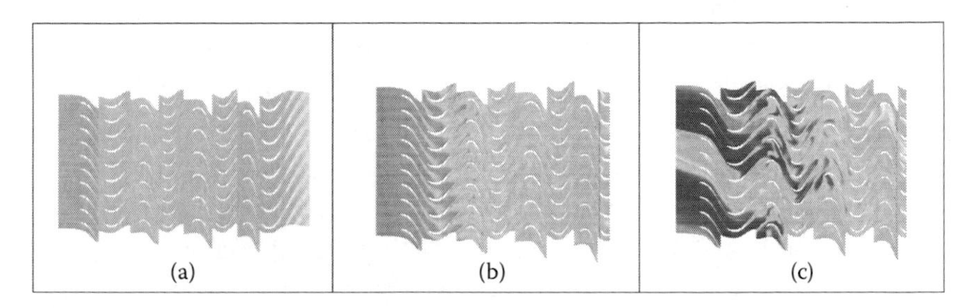

FIGURE 19.2 **(See color insert.)** Entropy simulation through multistage low-pressure turbine. (a) Steady uniform inlet. (b) Unsteady uniform inlet. (c) Unsteady including strut wake. The white crescents are cross-sections of the blades. The simulation looks at the flow between blades in what is referred to as a "blade passage."

stand-in data where internal, proprietary models would not be desirable under the terms of the grants' user agreements. Similarly, the nature of the studies would be of scientific interest, and preferably align with the DOE mission. One project perfectly fitting the target interests of the government programs emerged in modeling the freezing of water. Although such a phenomenon may seem very basic, the underlying mechanisms of ice formation is not well understood by science, and yet is of great interest for better designs of GE products placed in frigid environments. Along with its sheer scale, the pure scientific nature of this problem made it our case study of choice to follow.

DOE's Leadership supercomputers played a vital role in GE's future dedication to high-end modeling by vetting the feasibility of use and achieving never-before-possible results. Although government systems remain crucial to engineering and science problems of general interest, the most tangible outcome certifying the success of these collaborations with the government labs was the 2011 investment in an internally owned supercomputer dedicated to targeting GE's most proprietary data.

19.3 CASE STUDY: ANTI-ICING SURFACES

19.3.1 Project Background

For several years, GE Global Research has been developing anti-icing technologies to prevent the accumulation of ice on critical surfaces. "Anti-icing" can be used to describe several candidate approaches, including those that foster water droplets to bounce off a surface before they freeze, suppress the formation of ice nuclei, or facilitate easy shedding of ice in the event that it forms. The development of anti-icing surfaces has been a part of a larger effort to understand how nanotechnology affects macroscopic surface properties, including hydrophobicity, self-cleaning, thermal conductivity, and condensation. The motivation of this generalized approach to surface engineering has been opportunistic, the intent being that positive findings could impact many GE businesses including aviation, oil and gas, power and water, lighting and appliances.

Much of the initial activity was driven by experiment, including the development of textured hydrophobic plastic and metals [1–2]. GE Global Research developed many sophisticated experimental facilities for the anti-icing studies, including a custom-made ice tunnel and freezer fan for evaluation of the shear strength and centrifugal forces required to remove ice. In addition, GE Global Research developed an apparatus with a high-speed video camera to capture the dynamics of supercooled water droplets launched onto a cooled surface, including two infrared (IR) cameras to monitor the droplet temperature at the tip of the needle and at the surface on impact.

After several years of extensive lab study, molecular dynamics computer simulations were incorporated into the program as a modest supplemental effort. A typical rationale for modeling is that it is faster or cheaper than experimentation. However, neither argument could be made in this case. Molecular dynamics is a notoriously compute-hungry technique. Millions of CPU-hours have been used to model miniscule droplets for a fraction of a second. Although this author has personally observed billion-fold productivity gain in the 20 years since she wrote her first molecular dynamics code, said author concedes that the system sizes and timescales in the molecular dynamics world are miniscule relative to what is observed in the physical world.

To appreciate the relevant physical and temporal scales, consider that lab experiments contained moles* of atoms and observations over seconds or minutes. The state-of-the-art of molecular dynamics is "billion atom" and "milliseconds," and even then, usually not at the same time. To appreciate the timescale complexity required for modeling freezing at the molecular scale, each computational step simulates one femtosecond (10^{-15}) in the physical world. Thus, modeling even one microsecond in the real-world demands a sequence of a billion calculations, keeping in mind this is done for all relevant interactions between molecules in the

* A mole is a number so large (6×10^{23}) it is difficult to comprehend. For analogies, including a song, see http://molechemistry.info/.

simulation. It should be clear the sheer size of even a minimally significant simulation will therefore require a supercomputer.

What motivated the team's decision to incorporate molecular dynamics was the desire to better understand the fundamental science behind observed data. In particular, the experimental team observed that high contact angle surfaces could substantially delay the onset of freezing, confirming our original hypothesis that a lower contact area or hydrophobic surface can delay the formation of ice. However, we also observed that this benefit was temperature-dependent [3].

The assumption driving these experimental studies was that freezing starts somewhere at the water/surface interface. Thus, the premise is that modifying the surface can hinder the onset of freezing. When a cluster of molecules within the water droplet spontaneously arranges itself in an ice-like configuration (something like an iceberg) that is larger than a critical size, the ice front propagates through the system. Although the exact size of the critical nucleus depends on ambient conditions, it is likely that it contains only 100 or so molecules spanning just a couple nanometers. The molecules within the "iceberg" differ from their neighbors in only the slightest shift in relative position, making it extremely challenging to observe the process of critical nucleus formation, especially deep within a droplet. We identified the onset of freezing experimentally by using an IR camera to measure the macroscopic release of heat. While this method can identify *when* freezing occurs, it cannot identify *where* it occurs. Indeed, no experimental method has sufficient spatial resolution to determine the exact location of a critical nucleus spanning only a few nanometers across. Although computationally expensive and time-consuming, molecular dynamics can track the position of each water molecule and pinpoint when and where the freezing process starts. Moreover, the input parameters (e.g., ambient temperature, quench rate, heat-removal rate) can be probed in isolation, without parasitic variables such as humidity, wind, or dust interfering with the basic understanding.

19.3.2 Critical Role of the Greater HPC Community

Industrial researchers are driven to achieve the best possible result given demanding constraints on money and time. Given the considerable investment—and trust—already placed on experimental work, the simulation team needed to quickly demonstrate credibility and make an impact. Key to establishing credibility is to replicate existing experimental results, followed by making virtual predictions that can later be confirmed by experiment. We immediately started running modest simulations on our in-house Linux clusters. We gratefully partook of the robust, scalable open-source molecular dynamics codes developed by the greater community, such that our own expertise could be committed to defining the problem, analyzing the results, and applying the findings.

The simulation work especially took off after the team started gaining access to leadership-class supercomputers, in particular the Cray "Jaguar" and "Titan" located at Oak Ridge National Lab. Although continuing to run simulations on our local resources to make continuous progress, we applied for two Director's awards on "Jaguar" to execute jugular simulations on a much larger scale. After we gained sufficient confidence in the utility and scalability of our models, we proceeded to compete for a series of leadership-class compute-time awards, thus far winning two Department of Energy Advanced Scientific Computing Research (ASCR) Leadership Challenge Awards (2010, 2012) totaling 80 million CPU hours.

In addition to these industry-wide advances, we have benefitted tremendously from the personal contributions of several experts in their respective fields, including: Mike Brown (ORNL, GPU acceleration [4]), Art Voter (LANL, Parallel Replica Method [5]), Paul Crozier (Sandia, dynamic load balancing), Valeria Molinero (Utah, mW (molecular weight) water potential [6,7]), Mike Matheson (ORNL, visualization), and Aaron Keyes (Michigan, crystal pattern-matching). By staying on top of—and actively contributing toward—the state of the art in the field, we have already achieved ~200 × productivity increase relative to our original ALCC proposal written in 2010. In addition, we are optimistic that our present activity will lead to another 10–100 × acceleration in the observation of rare events, that is, nucleation.

Each one of these changes required a modification of the project plan, including an investment of time to incorporate these changes (for example, when we switched water potentials, we had to also develop a new water-substrate interaction potential because none existed in the literature) but these investments have more than paid off. Although we clocked in at 6 femtoseconds per 1024-CPU-second for a million-molecule droplet in 2011, we consistently achieved 2 picoseconds per 1024-CPU-second for a similar-sized droplet in 2013.

19.3.3 Opportunity in Cold Climate Wind Energy Generation

At over 850 tera-watt hours/year, wind energy presently represents 4% of the global electricity generation. Production capacity is expected to more than double by 2030. A sizable portion of this growth is projected to be in cold regions, which possess good wind conditions and low population density. At the same time, installation of wind turbines in these regions requires that effective ice-mitigation strategies be implemented. Up to 3%–10% of the energy generated by a wind turbine can be lost due to ice, as a result of the energy required to heat the blades or of shutdowns.

Even beyond this immediate context, our molecular dynamics study to develop anti-icing surfaces is proving to be interesting on many different fronts, including:

- Basic science—understanding the effect of freezing water, especially in the presence of surfaces, is not only a multidisciplinary engineering problem, it is also key to understanding some of the most basic questions pertaining to the existence and continuation of life. Because the mechanisms of freezing (and, by extension, the means by which to avoid freezing) are challenging to probe experimentally [8] there is tremendous opportunity for simulations to provide fundamental insight across many disciplines:
    - Ice formation is important for study of climate, geology, and biology.
    - In addition to surface interactions, chemical reactions in ice are also of great industrial interest and this work is a stepping stone toward such understanding.
- Computational—compute-hungry, Big Data problem requires novel acceleration and visualization/analysis schemes.
- Policy—clean energy generation is an imperative.
- Business—anti-icing surfaces can be a differentiator to expand into new markets.

Although the research goals of industrial researchers do not necessarily coincide with those working in academia or the national labs, the multi-faceted nature of this particular problem—including the richness of the basic scientific challenge—enables our group to contribute to (and not just benefit from) the dialogue at the very cutting edge of the HPC community.

In addition, the project has attracted significant attention in the trade press, including being the recipient of a 2013 IDC HPC Innovation Award, 2013 HPCwire Editors' Choice Award, and 2014 HPCwire "People to Watch" citation.

19.3.4 Overview of Molecular Dynamics Techniques

Molecular dynamics is one of the most active areas within the field of HPC, as evinced by a quick scan of awardees of leadership-class computing challenges, such as the Department of Energy ALCC and INCITE awards and the National Science Foundation PRAC award. On the other hand, molecular dynamics is a notoriously compute-intensive technique, as a given molecule (in principle) interacts with every other molecule within the system. If a pair-wise potential is used, the simulations scale as $O(n^2)$, assuming all of the interactions are included. Moreover, multiple floating-point operations are required to calculate inverse distance terms for the force and potential expressions. Although most modern molecular dynamics codes scale as $O(n)$ or $O(n \log n)$ [9], the reality is that the present day upper limit of molecular dynamics is about 1 billion molecules, which is about 15 orders of magnitude smaller than what one would typically consider a physical system. Moreover, along with protein folding, our topic of choice—spontaneous freezing—is considered a particularly slow molecular dynamics problem.

Although our team initially considered launching simulations on the order of a billion molecules, which would be unprecedented for a molecular dynamics study that probes freezing, we decided that one-of-a-kind "hero" simulations would prevent our modeling the wide range of parameters required to provide useful feedback to the experimental team. We decided that finite computational resources could be better spent distributed in the following manner, which in aggregate, makes for a leadership-class study:

- Large system—1 million molecules
- Long duration—up to a billion time steps
- Many initial conditions—six different surfaces, four different temperatures
- Many replicates—32–200 identical state points

There is a reason why many replicates are required. Spontaneous nucleation falls under the category of a stochastic process, where "rare" events are being monitored. Indeed, it is entirely possible that a critical nucleus may not form within a given droplet within a certain observation time. Thus, it is critical to implement several replicates under the same ambient conditions, such that the wide distribution of potential outcomes can be adequately captured.

In part because of the importance of the role of water in biological processes, dozens of water potentials have been developed over the past decades. The most popular water models have been based on point-charges, including 3-, 4-, and 5-point-charges. One of the key changes made on our project relative to our initial proposal was switching from a traditional 3-point-charge pair-wise interaction potential to a new 3-body potential that has recently been growing in popularity [6]. This required that we develop our own water/substrate interaction potential, as none existed in the literature. This also spurred us to develop hybrid CPU/GPU code for 3-body interaction potentials [4] such that the code could run effectively on the newly commissioned "Titan" supercomputer at Oak Ridge National Lab (Figures 19.3 and 19.4).

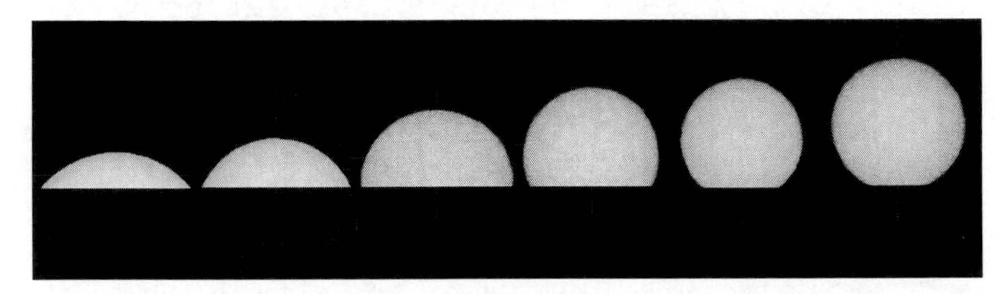

FIGURE 19.3 1 million molecule water droplets on 6 different surfaces, from low contact angle (left) to high contact angle (right). Contact angle corresponds to the degree to which water droplets are repelled from the surface.

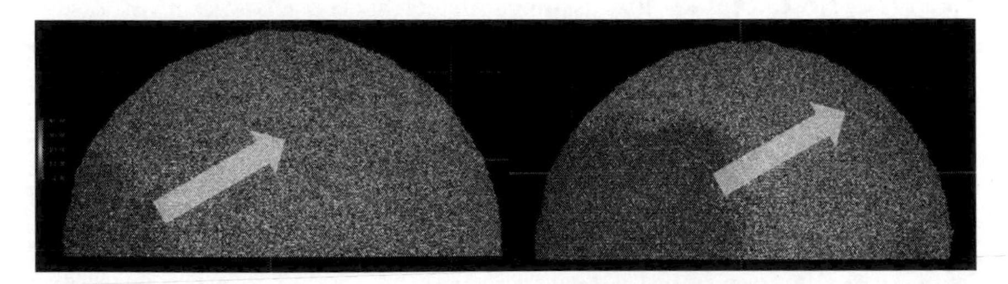

FIGURE 19.4 (**See color insert.**) Critical nucleus propagating through droplet, where droplet mobility is denoted by color (blue: low mobility, white: average mobility, red: high mobility). As liquid water turns to ice, latent heat is released, causing the molecules surrounding the nucleus to heat up.

19.3.5 Equally Important: Data Processing and Visualization

One of the legitimate concerns of the referees evaluating our proposals was that, even if the simulations could be successfully executed, would it be possible to extract meaningful information from the copious data? We estimate that we will have about 50 terabytes (TB) of data at the conclusion of our simulation activity. Our droplets contain 1 million molecules and we are trying to identify clusters of molecules on the order of 100 molecules. That means we are searching for a signal that affects only 0.01% of the total number of molecules, where the distinguishing characteristic of the relevant molecules is an almost imperceptible shift in position and angle. Moreover, we have no a priori knowledge of which of the molecules will contribute to the nucleus, nor do we have any a priori knowledge of where any of the molecules will be at any given time. This means we cannot simply scroll down a list of molecule index numbers and determine whether molecule 2 or molecule 455,328 is part of the nucleus. It is the slightest *relative* rearrangement of the molecules that needs to be tracked. Relative distance (pair-wise relationship) does not provide sufficient information; relative angles (3-body relationship) must also be calculated. Thus, detecting ice clusters is a process that requires the application of recursive pattern matching algorithms.

Even after the molecules within the ice cluster are properly tagged, effective visualization schemes must be used to understand when, where, and how the nucleus forms. This is again a domain where we have benefitted by rapid advances made in the field. When we first started visualizing million-molecule droplets in 2010, we found ourselves in a bind. If we ported the files onto our work laptops and visualized the files locally, the laptop would crash. However, if we tried to visualize remotely, the visualization screen would not respond for minutes after a command was sent. Now, thanks to advances in dedicated visualization nodes, we are able to directly visualize data from the Oak Ridge National Lab directories, without porting the files locally, in close-to-real time.

19.4 MOLECULAR DYNAMICS SCIENCE

19.4.1 What's Next

Of course, the critical question after all the simulations are complete is what technical insights have been gained. At this point, we know that our simulations have replicated the trends that have been observed in the lab. Namely, we have observed that surfaces that repel the adhesion of water (high contact angle surfaces) directly correlate to a delay in freezing.

Moreover, we have observed that this surface-mediated delay in freezing is temperature-dependent. More critically, we have observed that the nucleation mechanism is not the same across all operating conditions; this information can be used to help understand which anti-icing strategies are most likely to be effective under a given set of circumstances. These findings will be published in a peer-reviewed scientific journal after we do a complete analysis of all 50TB of simulation data.

Although we believe this aggregate study to be one of the most comprehensive of its kind, there is still much room for growth in the field before truly realistic simulations can be achieved. For example, we have not included the effect of wind, droplet impact velocity, dust, substrate conductivity, surface texture, surface charge, or flexibility of the substrate in our model; these factors are known to affect freezing behavior in real life but they add tremendous complexity to an already compute-intensive model. In addition, the simulation size and simulation duration are still orders of magnitude away from observed physical scales, requiring a leap from a billion molecules (10^9) to moles (10^23) and from milliseconds to minutes, hours, or even years. Advances in software/algorithms and hardware must occur in parallel to bridge this gap.

Finally, even if molecular dynamics simulations prove to be the key that opens doors to new anti-icing surface designs, tremendous additional investment is required to develop a surface that fulfills the requirements for manufacturability, durability, and cost. Although our simulations have already provided scientific insights that can help inform business decisions, the fact remains that the successful launch of any new product depends on an entire industrial ecosystem coming to fruition, including technical risk-reduction, harsh environment testing, manufacturing scale-up, supply-chain optimization, mobilization of commercial teams, regulatory management, installation/deployment, long-term maintenance and software, controls, and analytics.

We are emboldened by the technology developments that have been enabled by HPC and are optimistic that the power and utility of HPC will incorporate more aspects of the smart/advanced manufacturing in the future. We look forward to HPC helping to realize speedup not only within the engineering function, but in the entire new product introduction process.

19.4.2 Industrial HPC Applications Landscape and Impact

The GEnx aircraft engine powering Boeing's 787 Dreamliner includes advanced capabilities made possible through HPC modeling and simulation.

In particular, extremely complex simulations were conducted in designing the LPT (low-pressure turbine). Simultaneous simulation and analysis of all six stages of the LPT, a scale impossible prior to today's supercomputers, have the potential of improving fuel efficiency 1%–2%. That would equate to hundreds of millions of dollars in annual fuel savings across the aviation industry. Far greater efficiency improvements are being researched as well, and hinge on insights from studying the multiphysical system interaction of chemistry, acoustics, and fluid flows. The timescales of the dynamics and nature of these interactions cannot be physically measured. Only HPC simulation can advance the scientific understanding and engineering design.

Related programs must understand thermal modeling and structural mechanics of these systems, both for operations and maintenance. The systems employed in controlling the engine are increasingly complex and leveraging physics-based insights into performance and efficiency. Predictive health maintenance services employ models for reliability and part life prediction to manage costs and minimize downtime. This is increasingly important as markets expand globally. Process modeling for manufacturing and materials design now reach into HPC-scale platforms to improve throughput and yield.

For perspective, the current global annual fuel consumption in GE's turbines for aviation and power generation is approximately $200 billion. These highly engineered machines will rely on advanced computational modeling for efficiency improvements such as the GEnx LPT mentioned earlier. Each 1% reduction in fuel consumption means $2 billion in long-term savings and proportional reduction in greenhouse gas emissions.

19.4.3 Advanced Manufacturing, Big Data, and the Industrial Internet

As the GE Design Engineering research community pushes forward the state of their art, their flagship successes with supercomputing-fueled modeling and simulation embolden and challenge both their technical peers and the downstream engineers focusing on manufacturing, supply-chain logistics and MRO (maintenance, repair, and operations). A grand opportunity is unfolding to more intimately link the processes of product design with materials design with manufacturing and ultimately field services. For an original-equipment manufacturer (OEM) company like GE, the ability to leverage knowledge across these phases can offer tremendous value to its customers. The challenge is to match the model data and appropriate simulation timeframes at relevant fidelities between the end-to-end product phases. For example, a design engineer may try a hundred different ideas in the amount of time a manufacturing simulation can assess the feasibility of just one.

To truly harness the potential of computational and cyber-physical technologies, however, the workforce behind manufacturing and service will need to be as skilled working with data as they are today with raw materials like metals. They will need the skills of symbolic logic and procedural programming to converse with the enabling machinery, be it software systems, robotic controls, complex sensors, or additive manufacturing's 3D printers.[18] Computational abundance, scalable data analytics, agile methodologies, and the sophisticated interconnectivity of people, machines, and software through the Industrial Internet[19] underlie the realization of today's Advanced Manufacturing vision.

One significant feature of Tera-to-Petascale computational modeling and the Industrial Internet is the creation of *Big Data*. Through scalable analytics over increasingly vast data, new kinds of discoveries are enabled via data-intensive processing. The bottom-up pattern-seeking strategy of data mining and the top-down data generation from high-fidelity simulation share common challenges now targeted by rapidly evolving knowledge discovery tools. For example, the strategy of simulations being co-launched with data management and analysis software agents[20] may prove to be necessary for deriving practical result sets for many high-end problems. The data management and analysis burdens precipitating from such high-fidelity simulations will become more tenable as HPC in turn lends performance power to the emerging ecosystem of Big Data tools for sensing, acquisition, quality, classification, resolution, transmission, storage, analysis, aggregation, search, discovery and visualization of structured data.

The Industrial Internet is the result of a vision of creating wholly new kinds of experiences and systems enabled by an interconnectedness of machines, people, software, and data that was not previously possible. Such an infrastructure provides fertile soil for developing, validating, and tuning complex numerical models through connectivity to measured physical systems. The engineering community is beginning to see the potential for reducing costs associated with quantifying confidence and improving model precision and accuracy limit application of simulated systems.

19.5 CONCLUSION

Companies like General Electric that lead innovation in highly technical industries must keep pace with the engines of discovery and employ state-of-the-art tools to remain globally competitive. The impact of

Moore's Laws on both the raw capability of high-end systems and their affordability assure increasingly high-fidelity data will be generated from ever-increasing scales of simulation and modeling. Beyond the challenges of technical practice, there remain cultural, educational, and business model hurdles to overcome before fully employing the potential of high-performance simulation-based engineering and science. From the synthesis of previously distinct physical models across multiple disciplines to novel strategies for taming data management and analysis, high-end computing will be central to strategic change in how products and services are conceived, designed, created, distributed, and managed. However, the mere existence of these advanced technologies are insufficient for success, and as ever it will rely on the talents, skills, passion, and creativity of people in employing such tools, from shaping a lofty vision to carrying out detailed tasks in execution. The main difference will be the sophistication and significance of the computational system as a scientific instrument and analytical collaborator in the process.

REFERENCES

1. http://www.technologyreview.com/news/405378/super-repellent-plastic/.
2. http://www.technologyreview.com/news/410974/water-repelling-metals/.
3. Azar Alizadeh, Masako Yamada, Ri Li et al. Dynamics of ice nucleation on water repellent surfaces. *Langmuir* 2012, 28, 3180–3186.
4. Michael, B. W. and Masako Y. Implementing Molecular Dynamics on Hybrid High Performance Computers — Three-Body Potentials. *Computer Physics Communications*, 2013.
5. Voter, A. F. Parallel replica method for dynamics of infrequent events. *Physical Review B*, 1998, 57, 22, R13985–R13988.
6. Molinero, V. and Moore, E. B. Water Modeled as an Intermediate Element between Carbon and Silicon†. *The Journal of Physical Chemistry B*, 2008, 113, 13, 4008–4016.
7. Moore, E. B. and Molinero, V. Structural transformation in supercooled water controls the crystallization rate of ice. *Nature*, 2011, 479, 7374, 506–508.
8. Thorsten Bartels-Rausch. Chemistry: Ten Things We Need to Know About Ice and Snow. *Nature*, 2013. http://www.nature.com/nature/journal/v494/n7435/full/494027a.html.
9. Big O notation. http://en.wikipedia.org/wiki/Big_O_notation.

ENDNOTES

1. Founded in 1900, headquartered in Niskayuna, NY, and designated a National Historic Landmark in 1975.
2. http://en.wikipedia.org/wiki/Edisonian_approach - trial and error (or hunt and try).

3. Visit computerhistory.org for some nostalgic product brochures (http://bit.ly/1iQbBdt).
4. An influence on operating systems (UNIX)—jointly with MIT and Bell Labs (http://web.mit.edu/multics-history).
5. 1993 ACM Turing Award (Stearns and Hartmanis).
6. Fiduccia–Mattheyses algorithm (FM algorithm).
7. Lorenson and Cline—Marching cubes: A high resolution 3D surface construction algorithm.
8. Foundational textbook: Object-Oriented Modeling and Design (Rumbaugh, Blaha, Premerlani, Eddy, Lorenson).
9. GE employed 1980's HPC technologies from vendors including Cray, Convex, and Thinking Machines.
10. GE was among the earliest and largest adopters of Sun Microsystems workstations.
11. GE Aerospace sold to Martin Marietta in 1993, merged with Lockheed in 1995, forming Lockheed-Martin.
12. In modeling, first principals or "ab initio" refers to directly solving the fundamental physical mathematics of the system (rather than a model based on abstraction, approximation, or data correlation).
13. The modeling and simulation community owes much to the U.S. Department of Energy's advancement of algorithms, software, system infrastructure, and hardware in serving its mission to numerically model the safety and readiness of the nuclear arsenal under the Advanced Simulation and Computing (ASC) Campaign.
14. LLNL/IBM 2005 Gordon Bell Prize simulating metal solidification (http://1.usa.gov/1fUf9gZ).
15. Prof. Ken Jansen's Computational Fluid Dynamics code (PHASTA) porting and scaling well on IBM BG/L.
16. Weak Scaling refers to efficiently increasing the size of the problem able to be solved in proportion to the number of processors used, while Strong Scaling is reducing the runtime of a constant-sized problem in proportion to the number of processors on which it is run.
17. "Overcoming the Turbulent-Mixing Characterization Barrier to Green Energy and Propulsion Systems," Gupta, Laskowski, Shen, and Shieh, General Electric Global Research, Argonne National Laboratory IBM Blue Gene/P (Intrepid).
18. Discussing skill demand: http://www.IndustrialInternet.com/blog/stemming-the-tide-of-outsourced-employment/.
19. GE has proposed the name "Industrial Internet" for a vision of technology advancement through the growing pervasiveness of sensor instrumentation, ubiquity of capable computing, and pervasive connectedness of minds and machines.
20. In-Situ Analytics describes such an approach. For example, see http://exactcodesign.org.

CHAPTER 20

An Italian Twist on an America's Cup Sailboat Hull Design

Raffaele Ponzini and Andrea Penza

CONTENTS

20.1 INTRODUCTION

To win the America's Cup [1], one must defeat the other teams not only in regattas but also in design studies, a strongly technological challenge where a major role is covered by computational fluid dynamics (CFD). As is common in most computer-aided engineering (CAE) industrial applications, software licenses are a major cost item; moreover, it is well known that in-silico data acquisition (i.e., data obtained from computational models) is becoming the elective environment for the design development process, involving multiphysics CAE tools and a dramatic increase in requested high-performance computing (HPC) resources. As a byproduct, budgets related to "computing" activities are increasingly at risk of becoming the real bottleneck in the technological challenge. Open-source CFD software might become a key point to overcome such a bottleneck; nevertheless, large campaigns are required to validate digital models against physical models to ensure reliability for industrial-use cases of interest. The problem is complex in its entirety and it is crucial to have access to advanced computing resources that are able to process the large amount of data produced by such design studies considered as a whole and by each single numerical simulation. For this purpose, the design team of the Luna Rossa Challenge [2], besides taking advantage of large amount of HPC resources for design data production, both in terms of computing and remote visualization, undertook with CINECA in 2012, a 12-month feasibility study evaluating the OpenFOAM (Open Source Field Operation and Manipulation) toolbox [3].

20.2 ADOPTION OF HPC: INSIGHTS INTO PORTABILITY AND SCALABILITY

The validation project presented herein has been designed to study both state-of-the-art AC72 hulls, those that ended up competing in the San Francisco Bay waters in August 2013, and well-known benchmark hulls, fully validated through computational tools and towing tank experiments. To provide fruitful answers, we tested both the accuracy of the open-source solution against other commercial CFD codes and its computational efficiency concerning the time-to-solution at different degrees of parallelism.

As a first attempt, we studied the hull hydrodynamic of a reference hull known in literature as DTMB–5415, a model conceived by the U.S. David Taylor Model Basin (DTMB) in the early 1980s as a preliminary design for a surface combatant ship with a sonar dome bow and transom stern. For intellectual property reasons, all the results detailed in this article concern this hull type. Suffice to say that we found that the same general conclusions may be applied also to the racing hulls in design.

The Navier–Stokes physics equations that govern the motion of fluids are particularly complex to solve. Only in very rare cases, and with simplified geometry and conditions, is it possible to obtain an analytical solution. Therefore, digital numerical methods are considered the only feasible way to solve such complex equations in real-life geometry and flow conditions. The library used for this validation work, OpenFOAM, is based on breaking down, or discretization, of the problem obtained through the so-called finite-volume method. The equations are strongly nonlinear and coupled, which makes the search for a solution even more complex, even when approximated by numerical schemes. Furthermore, the fluid-dynamic phenomena in the marine environment exhibit random turbulent behavior, and this requires additional modifications of the equations. When dealing with complex three-dimensional (3D) geometries, it is as yet impossible to perform a calculation of the flow field in the regime of turbulence solving the equations in a direct way (DNS [direct numerical simulation]) without the use of a simplified turbulence model. We, therefore, need to reduce the complexity using the so-called RANS (Reynolds-averaged Navier–Stokes) equations. Given that the nature of the problem is also multiphase (as we need to study and capture with accuracy the interface between air and water) it is necessary to add a further equation to the usual set of Navier–Stokes equations; namely, the equation for the mass fraction, in agreement with the so-called volume of fluid (VOF) method. The general idea of this method is to have an additional variable that for each cell of the computational grid takes the value 1 in the presence of water, whereas the value 0 is associated with the presence of air [4,5].

OpenFOAM is an open-source library written in C++ that allows simulation of physical and mechanical behaviors, including those of the computational fluid dynamics. The toolbox includes not only numerical solvers but also tools for the pre- and postprocessing tasks required for a complete treatment of a CFD analysis. In particular, there is a generator of unstructured grid computational meshes, snappyHexMesh. OpenFOAM allows the user to perform simulations in parallel and to exploit HPC architectures with high efficiency. For the multiphase flow analysis described, we used the interFoam solver, and, therefore, the simulation was performed at 0 degrees of freedom (boat blocked). This seemed reasonable as a first approach of the multiphase analysis; however, OpenFOAM also provides the ability to perform similar simulations by releasing all or only some constraints of the boat, using the interDyMFoam solver. It was decided to perform calculations for Froude numbers (a dimensionless number related to the speed of the boat) in a range between 0.38 and 0.55, a satisfactory range for a sailing boat hull study. The value of the resistance produced by the hull surface advancing into the flow field was considered together with the contributions of the components of the pressure drag and viscous resistance, or fluid drag. For a qualitative analysis of validation, we used an output parameter typically used in the field of marine CFD to extract information about the wave form produced by the hull and consequently a reliable estimate of the efforts acting on the boat.

20.3 HPC PERFORMANCE EVALUATION AND BENEFITS

The primary HPC infrastructure used for this study was the Lagrange cluster at the CINECA Milan site, comprising a parallel computing system Hewlett-Packard C7000® with 336 nodes BL460c with Intel® Xeon® processor X5660 (exa-core), operating system Red Hat Linux 6.0, 24 GB RAM/node, and a low-latency network interconnection InfiniBand® QDR. The system was ranked number 210 on the TOP500 List of supercomputers in June 2010.

The quantitative analysis of the results in comparison with the references available was satisfactory with the percentage differences being less than 5% in the case of viscous resistance, and even less than 0.4% in the case of the resistance related to pressure contribution (see Figure 20.1). In the perspective of an industrial application of this workflow, we tried to reduce the computational cost by performing simulations only on half hull, exploiting the symmetry of the problem. This was proven to be successfully achievable and the results were consistent with those obtained for the entire hull allowing a total saving in computational time of about 43%. Scalability tests (Figure 20.2) confirmed the possibility to compress the computational time with excellent efficiency for high number of computational cores allowing for a full exploitation of HPC platforms and resources.

20.4 FUTURE DEVELOPMENTS

The work methodology used proved to be winning and showed that the open-source software has the potential to be exploited in industrial fields of applications, even in very competitive ones. This capability, coupled with the ease in automation of the overall obtained workflow and with the availability of resources of an advanced computing center, can open many roads in the improvement of the design process related to racing sailing boats. Thanks to the continuous and rapid development of information technology, today it is even possible to improve the design process by directly coupling CFD datasets with a so-called velocity-prediction-program (VPP) code that calculates and provides boat performances at different sailing conditions in real-time. In this regard, the performance of super-linear scalability testing showed herein are very encouraging. Also, being open-source code, the possibility of having open access to the software itself allows the user to customize the working procedure to their needs, an absolutely crucial possibility in the perspective of an industrialization process. For these reasons, we consider the software OpenFOAM a cost–effective tool in CFD applications based on HPC infrastructures.

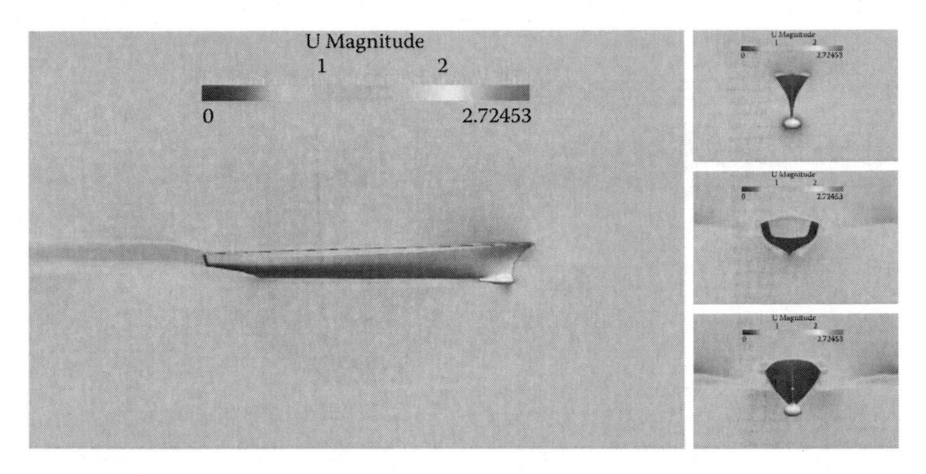

FIGURE 20.1 Symmetry well caught by the solver for the velocity field computation on the DTMB 5415 test case.

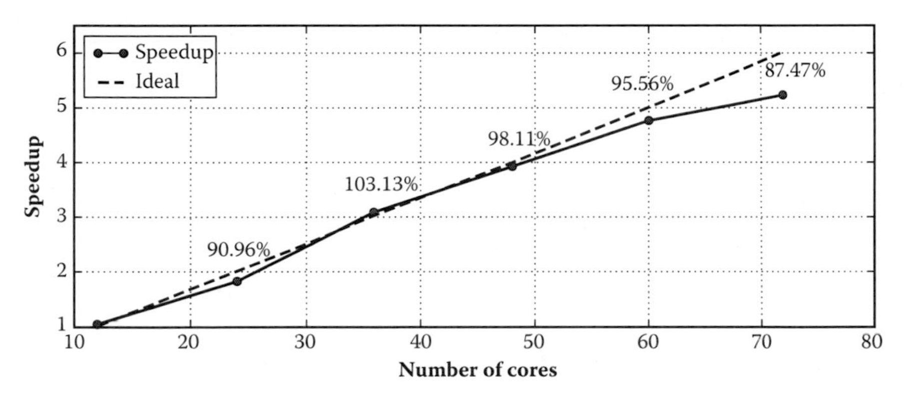

FIGURE 20.2 Scalability test for the DTMB 5415 hull under captive case at Froude number 0.38 and mesh size of about 2 million cells. Superlinearity caught at 36 cores for this mesh size problem.

REFERENCES

1. America's Cup, http://www.americascup.com.
2. Luna Rossa Challenge 2013, http://lunarossa-challenge-2013.americascup.com.
3. OpenFOAM: The Open Source Computational Fluid Dynamics (CFD) Toolbox, http://www.openfoam.com.
4. Y. M. Ahmed, "Numerical simulation for the free surface flow around a complex ship hull form at different Froude numbers," *Alexandria Engineering Journal*, vol. 50, pp. 229–235, 2011.
5. P. M. Carrica, R. V. Wilson, F. Stern, "Unsteady RANS simulation of the ship forward speed diffraction problem," *Computers & Fluids*, vol. 35, pp. 545–570, 2006.

CHAPTER 21

NDEMC Midwest Pilot in the United States of America

Merle Giles and Cynthia McIntyre

CONTENTS

21.1 INTRODUCTION

The National Digital Engineering and Manufacturing Consortium (NDEMC), a public–private partnership, was founded in 2010 by three Midwest university supercomputer centers, four FORTUNE100® manufacturers, the National Center for Manufacturing Sciences, and the U.S. Council on Competitiveness. John Deere, General Electric (GE), Lockheed Martin, and Procter & Gamble (P&G) each invested $500,000 of cash and in-kind services, matched equally by the U.S. Economic Development Administration (EDA). Small and medium manufacturers (SMMs) were chosen by each original equipment manufacturer (OEM) to participate in advanced digital modeling and simulation at no cost. The pilot lasted approximately 30 months.

NDEMC STAKEHOLDERS

OEMs ($2 million)
Deere & Company
General Electric
Lockheed Martin
Procter & Gamble

Solution Partners
NCSA (National Center for Supercomputing Applications, University of Illinois at Urbana-Champaign)
OSC (Ohio Supercomputer Center, The Ohio State University, Columbus, Indiana)
NCMS (National Center for Manufacturing Sciences, Ann Arbor, Michigan)
Purdue University, Indiana

Non-Profit Fiduciary
U.S. Council on Competitiveness

Federal Government ($2 million)
Economic Development Administration
White House (Office of Science and Technology Policy, Federal CTO)

State Governments
State of Ohio ($1 million)
State of Indiana ($150,000)

Other Signees
NSF (National Science Foundation)
NSF (National Science Foundation)
DOE (Department of Energy)
NIST (National Institute of Standards and Technology)
NASA (National Aeronautic and Space Administration)

Initially, two suppliers per OEM whose advanced modeling and simulation capabilities ranged from none to modest were chosen. In each case, these companies were using two-dimensional (2D) geometry CAD (computer-aided design) packages, whereas some companies were using 3D modeling applications on desktop computers. None were using supercomputers. The difference between 2D and 3D can be described as the comparison between a traditional blueprint and a 3D rotating image of a part that fits in an assembly (as in the *Iron Man* movies).[1] Advanced simulations explore structures using finite-element analysis (FEA), computational fluid dynamics (CFD), thermal progression (such as molten steel as it cools), and other physical domains, either separately or together. The most advanced simulations, known as multiphysics simulations, analyze two or more physical domains simultaneously.

The initial scope for NDEMC was to concentrate on 3D simulation training that used well-known commercial off-the-shelf (COTS) codes.

Structures and fluids are the two fundamental engineering domains with a handful of commercial codes from independent software vendors (ISVs) that supercomputer centers as well as small manufacturers would be familiar with, such as ANSYS Mechanical, ANSYS Fluent, Simulia Abaqus, and CD-adapco's Star-CCM+. With permission from the ISVs, benchmarks and analyses were conducted on university supercomputers. Performance improvements were quickly achieved using more processing power than was available on desktop computers. Structural inputs were evaluated, specific physics was verified, and expert analyses were made of the simulation techniques.

21.2 NDEMC CASE 1: JOHN DEERE AND ADAMS THERMAL

One of John Deere's suppliers is Adams Thermal Systems, a charge air cooler (CAC, or radiator) manufacturer in South Dakota. John Deere's desire to engage Adams Thermal was due in large part to the impact the radiator has on the efficiency of (and emissions from) an internal combustion engine. As one can imagine, the radiator on a tractor is large, and design decisions impact the placement of hoses for fluid inputs and outputs, the speed of fluid transport, the angles of turns, widths of openings, density of cooling fins, placement of welds, and more. Better-informed initial design choices produce long-lasting parts and assemblies and more efficient fuel consumption.

Digital design of experiments, or experimental design, is the design of any information-gathering exercise where variation is present. Iterative designs will help an engineer optimize a design, and running multiples of these designs are aided significantly by supercomputing in two ways: (1) running on a larger number of processors and (2) running multiple simulations at the same time. The typical platform for digital design is a computer workstation with 4–8 cores with one or two CPUs (central processing units). The software licensing for these applications is typically charged by the core, which is the smallest unit of processing, and the licenses are often calculated based on annual use. These applications are powerful, and their licenses can easily cost 10 to 30 times more than a desktop computer.

Adams Thermal would run a single CFD simulation on the entire computer for 3 days. NCSA's first task was to take the same input file used on the workstation and run it on as many cores as possible on a supercomputer, using the identical CFD application. Immediate improvements were achieved with this effort, running efficiently up to 192 cores from the previous eight, reducing runtime from 3 days to 3 hours. In essence, the simulation easily ran on 24 computers instead of one. Success!

The 3-hour runtime created an opportunity for changes in workflow at the company, promising as many as 12 times as many runs in the original 3-day runtime. These extra runs would allow the designer to assess iterations of the initial design, or a suite of options rather than a single design.

Analyzing the costs for this computing is straightforward, as costs at NCSA and other supercomputer centers are typically quoted by the core-hour. Running on 192 cores for 3 hours converts to 576 core-hours, costing less than $125. Predicting the cost of the software licenses for 192 cores is not straightforward, as ISVs typically do not sell to companies that have access to supercomputers. This is changing, however, and the pricing models are beginning to take into consideration a sliding scale for core-hour pricing. Closer cooperation between supercomputer centers and ISVs is expected to provide promising performance increases for SMMs.

All companies must get product out the door on time, so engineering workflow is central to meeting customer design targets. As no manufacturer has unlimited time to explore optimal designs, trade-offs are introduced and must be assessed. The use of coarser models with less realism, which is typically what happens on a desktop, can increase the sheer number of iterations. Higher-fidelity models could be produced that would add more design realism, but they would take much longer to run on a desktop. Alternatively, a supercomputer could be used to add significantly more realism in a timeframe that would fit within the design workflow allotted.

Adams Thermal opted to pursue more realistic models, prompting the following issues:

1. Costs for computing and licensing would need to be quantified, and regular access to HPC resources assured for this to become a repeatable workflow.

2. The vision of more realism yet could be achieved by coupling the fluid and structural physics models. Adding thermal assessment and metal fatigue analysis, however, created an extremely complex

multiphysics challenge that was determined to be beyond the current capability of the software applications.

Unleashing a vision is a ton of fun, and the promise of unlimited compute power can do strange things to normal people. As Adams, Deere, and NCSA continued to analyze the complexity of the CAC, it became evident that the ultimate goal would be to simulate the entire assembly (Figure 21.1). Structural parts included the frame and the internal metallic fins. Fluid dynamics would analyze the flow of water throughout the assembly, thermal progression analysis would help find hot spots, and proper joint fatigue analysis would inform engineers in ways that would improve life of the assembly and reduce costs.

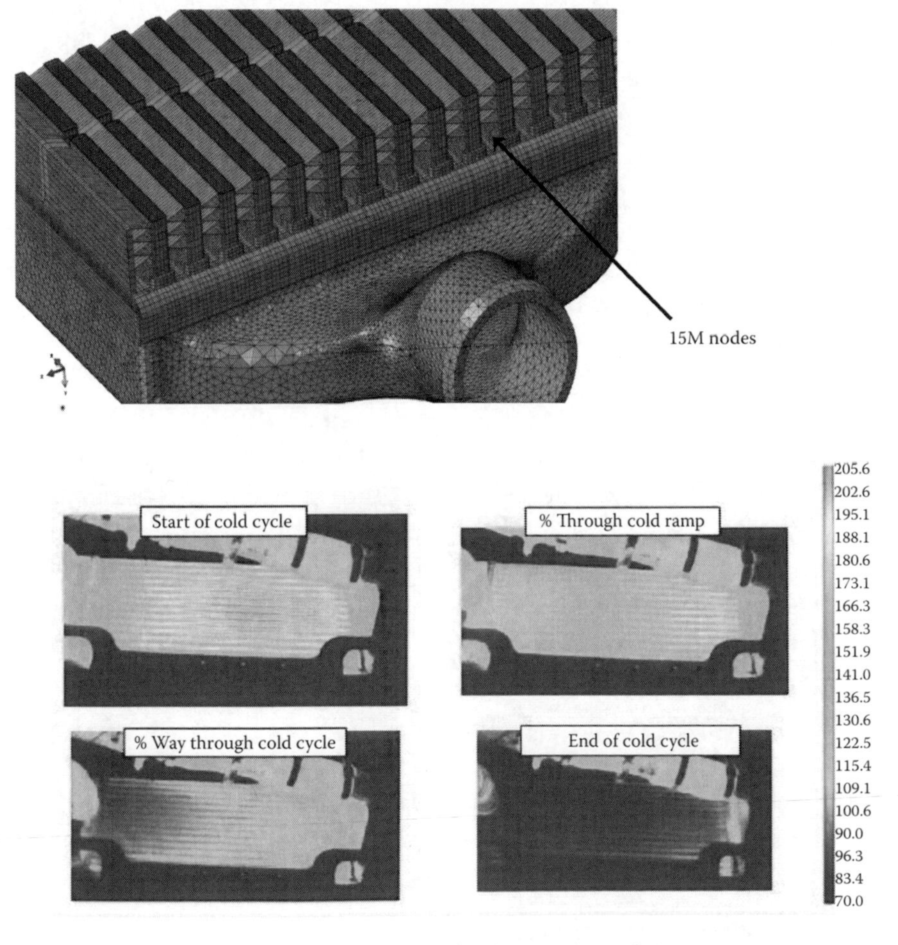

FIGURE 21.1 Multiphysics simulation of a charge air cooler.

The objective was to study the fatigue life of the CAC due to thermal stresses. A three-step sequential simulation was performed:

1. CFD of turbulent fluid flow through the CAC coupled with advective heat transfer (HT) provided thermal boundary conditions for structural analysis.
2. Structural analysis of the thermomechanical parts provides transient thermal stresses in the solid part during the thermal cycle for fatigue analysis.
3. Fatigue Model uses history of thermal stresses and estimates the cycle life at critical points.

Multiple physical analyses, or multiphysics, require the coupling of domain-specific codes to run together in time steps. Programmers must know enough about each code to link them to these time steps so they can be run across multiple CPUs simultaneously. Multiphysics is quite promising if a framework exists to link both fluid and structural codes. However, coupling domain codes from one ISV to the code of another is problematic without full cooperation from each ISV. ANSYS claims multiphysics between its own structural and fluids codes, and CD-adapco is in agreement to share with Simulia's Abaqus structural code.

Chapter 1 describes how Ford Motor Company used 20,000 structural elements for FEA designs in 1985. Today's highly advanced FEA designs will use 30 million elements. The Adams CAC model was estimated to require 100 million elements and a full man-month or two to develop such a sophisticated model. The complexity driven by this specific problem was extreme and not previously accomplished using commercial codes. NCSA and Deere's internal engineering team ultimately developed a reduced-order model and created a non-numerical method to add thermal values to the analysis, an effort beyond the scope of the NDEMC Midwest Pilot.

Lessons learned:

1. Funding for extreme physics-based modeling and simulation does not currently exist from science and engineering funding organizations.
2. One answer was received to the provocative question of what one would do with infinite compute power.

3. No one—absolutely no one—expected such an extreme engineering design challenge from a small company in South Dakota.
4. A recipe for repeatability is uncertain, but currently requires a reduced-order, less realistic model to meet cost and workflow demands.

21.3 NDEMC CASE STUDY 2: JOHN DEERE AND ROSENBOOM MACHINE AND TOOL

A family-owned John Deere supplier in Iowa designs and produces hydraulic cylinders (Figure 21.2) and machine parts for a variety of markets. Rosenboom engineers had no experience with 3D digital design simulation software, yet wanted to make more informed decisions about multiple design options within the typical single week allocated

FIGURE 21.2 A stack of hydraulic cylinders. (Photo credit: www.rosenboom.com).

for bidding for customer business. The initial strategy was to train the company engineers on CFD tools so they could more quickly and accurately analyze the behavior of hydraulic fluid in a cylinder.

NCSA hosted a 2-day training workshop in Champaign, Illinois for all NDEMC participants on a wintry day in January 2012, and several Rosenboom engineers attended. Fundamental aspects of CFD were taught using ANSYS Fluent, and FEA simulation was taught using Simulia Abaqus. All participants accessed NCSA's industrial supercomputer, iForge, in real-time.

Rosenboom engineers were thrilled with what the new CFD tools could do for them, quickly proving their return on investment (ROI) of time by using CFD to reduce 6–10 potential cylinder designs to a single optimal design with a measurable reduction in uncertainty. Rosenboom increased export sales by $7 million and hired 150 new workers during the project.

Lessons learned:

1. The highest ROI for training was using CFD.
2. Rosenboom engineers and management achieved unexpected benefits from learning FEA, as the ultimate goal of design was to reduce failures in the cylinder caps, which are welded to the cylinder body.
3. Student engineers with no previous experience with digital 3D design tools quickly became more productive.
4. The "obvious" choice to learn CFD did not achieve the ultimate goal of reducing fatigue on a welded cylinder cap. In ways very similar to Adams Thermal's desire to assess fatigue caused by thermal stresses, multiple physics analyses are required.
5. Understanding the engineering goals of the customer is paramount when introducing new tools.

21.4 NDEMC LESSONS FOR JOHN DEERE

Several OEM lessons should be noted from the NDEMC pilot:

1. The collaborative investment model between government and industry inspired behavior that would not have occurred by either side alone.
2. John Deere used the NDEMC project to survey its suppliers to determine the extent of advanced modeling and simulation being

conducted, asking for specific use of commercial, FEA, and CFD codes as determined by the initial NDEMC scope.

3. Deere had no deep insight into the extent of advanced modeling and simulation conducted in its supply chain.
4. Deere's existing partnership with NCSA's Private Sector Program provided a foundation of trust in technical capabilities that would last throughout the project and beyond.
5. Midway through the project, Deere management publicly stated that they had already achieved an ROI on their $500,000 cash and in-kind investment.
6. SMMs could not have achieved as much success as they did without hands-on involvement from Deere.
7. Support from Deere's CEO Sam Allen cleared the way for high attention within the company to this high-visibility project.

21.5 NDEMC CASE 3: GE, ESI, AND AN UNNAMED SUPPLIER

Success came in yet another unexpected manner with the GE-funded project. The chosen supplier manufactured wind turbine blades of significant length, and the project was to adopt 3D modeling of the entire turbine blade (Figure 21.3) rather than a fractional slice of the blade. ESI Group[2], founded in France in 1973, provided a CFD application, PAM-RTM, tuned for 2.5D on a workstation computer. GE's desire was to have its supplier adopt advanced 3D modeling and simulation. The supplier had no previous knowledge of 3D digital modeling.

The NDEMC funding agreement provided that no capital outlay would be required by the SMM other than in-kind time and attention. After training and engaging both the supplier and the ISV, the supplier opted not to move forward and ultimately withdrew from the NDEMC Midwest Pilot.

Failure often induces unintended consequences, and in this case it certainly did. The team's choice to move forward was as follows: improve the ESI code to full 3D capability or introduce another 3D application. Since the NDEMC modus operandi was to meet the customer on their terms, the team worked closely with the ESI software development team to improve code performance. The code had severe limitations in an HPC

FIGURE 21.3 A completed wind turbine blade on the factory floor.

environment because it had no message-passing interface (MPI) capability to run outside a single node. The NCSA team provided code development expertise to the ESI team directly, ultimately helping them adopt scalable multinode capabilities. Multinode capability was essential to adding sufficient realism to jump from 2.5D to full 3D simulations.

Lessons learned:

1. All SMMs are not created equally, and should not be expected to equally adopt advanced tools.
2. Rapid advances in 3D modeling at OEMs will put immense pressure on SMMs to adopt these digital tools. SMMs that choose to adopt will win. SMMs that choose the path of inertia will lose, because OEMs will not settle for yesterday's performance in today's competitive marketplace.
3. Not all engineers appreciate the benefits of HPC, yet once exposed to systems and expertise, most engineers can be expected to become rapid adopters.
4. OEMs have an inherent conflict of interest if they attempt to do too much for a single supplier.

5. Innovation is often a bottom-up process, but OEM knowledge and expertise is invaluable to all parties.
6. ISVs can benefit immensely from expert development teams at supercomputer centers.
7. Business models and cultural inertia are barriers to changing behavior and priorities.
8. ESI can become the hero in its PAM-RTM application niche if it collaborates with HPC centers and aggressively competes against non-MPI application vendors.

21.6 NDEMC CASE 4: JECO PLASTIC PRODUCTS, LLC

Jeco Plastics[3] is a small, custom-mold manufacturer of large, complex, high-tolerance products in Indianapolis, Indiana. The project chosen for NDEMC was to predict the performance of a plastic pallet (Figure 21.4) that was customized for Volkswagen, a large automotive OEM in Germany. The OEM was requesting last-minute design changes to an existing product. Company engineers had no previous experience with advanced digital modeling and simulation and were relying on tedious trial-and-error physical design and testing. High-ranking executives at the OEM suggested that Jeco adopt advanced digital tools.

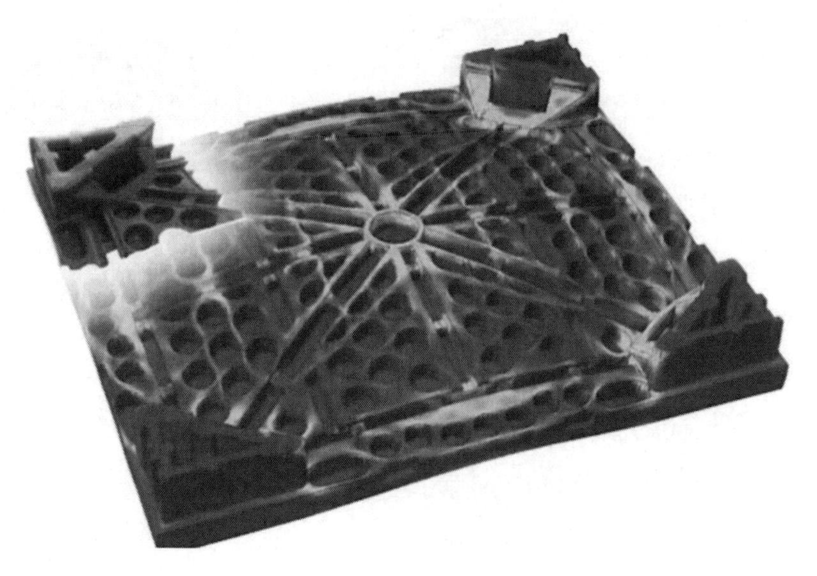

FIGURE 21.4 3D virtual image of a Jeco plastic pallet.

The technical challenge was to simulate complex, high-tolerance designs in inhomogeneous anisotropic materials. Anisotropy is a material's directional dependence of a physical property, such as the strength of wood when splitting along or against the grain. Jeco understood that the relatively small cosmetic alteration required by their client could potentially affect critical specifications for deflection, yet to secure the rather large contract they had to rapidly analyze the effect before making expensive, irreversible tool changes in the production process. NDEMC facilitated Jeco's access to expert staff at Purdue University and Ohio Supercomputer Center, who recommended the use of Abaqus[4] for FEA.

With the expertise and access to HPC-enabled advanced modeling and simulation, Jeco completed the technical analysis in time, securing a multi-year contract with annual orders of $2.5 million for the next 5–10 years. Fifteen jobs were added at the company following an initial capital investment of $500,000.

The Lessons learned were:

1. Jeco's demonstrated experience with advanced modeling and simulation during the NDEMC Midwest Pilot was instrumental in helping the company secure additional projects with aerospace and automotive customers.

2. Jeco subsequently received a lucrative order from NASA based on its new ability to design and manufacture products in layered anisotropic materials with continuous internal fiber reinforcement.

3. Advanced modeling and simulation has become a vital resource in Jeco's product development process.

21.7 NDEMC OVERALL LESSONS LEARNED

1. Train early and often.

2. Expect big things in small packages.

3. Do engage the OEM whenever possible.

4. Do invite the ISVs to engage technically.

5. Do treat the ISVs generously, as they will become either the bottleneck or the hero.

6. Do not immediately suggest changes in software applications to a customer, as there is neither time nor appetite to learn new tools in the initial engagements. Once a customer climbs the learning curve, more advanced software may be needed and desired.
7. Technical consulting staff must speak the languages of physics and engineering, not just HPC and CPUs.
8. The learning curve is steep when climbing alone. Always get professional help.
9. Companies need a trusted third-party advocate to guide them through the maze of software applications.
10. It is incredibly easy to get out of scope. NDEMC had wide measures of freedom in its Midwest Pilot, but care must be taken to properly fund more complex projects in a continuing organization.
11. More OEM/SMM collaboration is needed. The manufacturing supply chain cannot survive without it.
12. OEM technical leaders need to engage with suppliers. Procurement processes will not change without clear technical insight into supplier capabilities.
13. OEMs are not asking suppliers about their digital capabilities, but they should.
14. Development processes specific to suppliers' products drive the need for customized engagements, which are difficult to template and make repeatable.
15. The gap between OEM and SMM technical capability is widening. Some suppliers will be left behind, but OEMs need to engage with suppliers in new ways or they will risk losing all suppliers in certain niches.
16. Private sector investment is crucial to success. Without it, a government struggles to change behavior.
17. A "build-it-and-they-will-come" model will not work. Outreach needs to be on SMM terms with "boots on the ground."
18. Access to commercially viable supercomputing is crucial, as detailed proofs of concept are needed for rapid adoption.

19. The demand for subsidized services is endless. SMMs cannot afford to adopt advanced MS&A on their own, but with proper assessment of ROI, changes in behavior and SMM investment can be achieved. A key step in adoption is to reduce the uncertainty of the investment in digital tools, computing, and training.
20. Fundamental changes in costs for physical prototyping are required to adequately fund new digital tools and processes. If physical costs are 70% or 80% of R&D, they need to flip to 30% or 20% to make room for digital adoption. Alternatively, reductions in physical prototyping can conceivably fund investments in digital processes.
21. Strategic decisions must be made about subsidies to make SMM adoption of digital tools a lifestyle rather than a stunt.
22. Government-funded pilot programs are not designed to last forever.
23. Governance in a public–private partnership matters. Board members were industrial investors and the fiduciary U.S. Council on Competitiveness. Not represented on the board was any member of government, yet the EDA was a 50% investor. This was a mistake.
24. While all education and training may be local, collaboration is required, as not all HPC centers have expertise in advanced CAE modeling and simulation. E-learning is promising and must be attempted.
25. HPC center competition is unhealthy, because the real competition is not between machine providers. Expertise matters and there is not enough time and money to train all people equally.
26. OSC, Purdue, and NCSA learned to collaborate. Others can do so, too.
27. HPC centers must have the expertise and ability to manage commercial licenses; without it, all industrial engagement with SMMs will be "information only."

21.8 NDEMC OUTCOMES

The NDEMC team accomplished the following throughout the course of the project:[5]

1. Twenty SMMs were involved on more than a dozen projects.
2. Case studies were created for 10 of the demonstration projects.

3. Eight educational training videos were produced.
4. Several custom apps were created to simplify use of advanced modeling and simulation.
5. Based on workflow contributed by P&G, one app was specifically created for the simulation of fluid flows through a manifold. The Manifold Flow Predictor (MFP) accepts a CAD file and adds OpenFOAM CFD analysis and ParaView visualization. The MFP is hosted at Purdue University's ManufacturingHUB.org.
6. SMMs generated new business in excess of $25 million.
7. Related capital investment of $600,000 led to the creation of 160 new jobs.

ENDNOTES

1. http://www.youtube.com/watch?v=MqRVIEEp_AM.
2. https://www.esi-group.com/company/.
3. Council on Competitiveness Case Study, NDEMC Helps Jeco to Exceed Growth and Financial Expectations, ©2012.
4. Abaqus is owned by Simulia, a division of Dassault Systèmes in France.
5. NDEMC Final Report, August 29, 2013, www.ndemc.org.

CHAPTER 22

Geophysical Imaging in High-Performance Computing Platforms in Spain

Mauricio Hanzich, Josep de la Puente, and Francisco Ortigosa

CONTENTS

22.1 COMPANY'S TRADITION FOR INNOVATION

Repsol, a world top 50 energy company, has a long history of research and innovation. The Repsol Technology Centre (CTR) is a major R&D center in specialties such as petrochemistry and energy efficiency. Projects such as Kaleidoscope (exploration geophysics), Sherlock (reservoir characterization), or Repsol's prominent role in providing racing fuels and lubricants for World-Championship winning motorcycles have been renowned internationally. More than 400 researchers work at the CTR to improve product performance and environmental impact, thus "redefining the world energy model by applying new technologies that help with inventing the future."

22.1.1 Company and High-Performance Computing

A milestone for Repsol has been the Kaleidoscope Project.* The project started in 2007, uniting cutting-edge science and technology with the goal of boosting the finding of gas and petroleum deposits while reducing its environmental footprint. This has been possible thanks to exploiting the latest generations of computer chips together with state-of-the-art mathematical algorithms, which create accurate seismic images of the subsoil. These images enable us to "see" the earth's interior and search for hydrocarbons without drilling through the earth's crust.

This proprietary Repsol technology can process seismic data 15 times faster than other companies in the sector and greatly improves the chance of finding oil and gas thousands of meters below the subsoil. This technology has positioned Repsol at the forefront in the field of exploration of complex areas containing large undiscovered reservoirs, such as the Gulf of Mexico or the Brazilian shelf, where an estimated more than 100 billion barrels of oil lie undiscovered. The Kaleidoscope Project has since been named one of the five most innovative projects in the world by the American Institute of Electrical and Electronics Engineers. The *Computer World* magazine also awarded Repsol for this project, which has been a finalist in the Annual Creativity in Electronics Prizes and the Innovation Prize awarded by Petroleum Economist magazine.

22.1.2 Interaction with the National High-Performance Computing Ecosystem

The key to the success of the Kaleidoscope Project has been relying on some of the best geophysicists, mathematicians, geologists, and engineers from both Spain and the United States. This included researchers from Repsol as well as 3DGeo, a top geophysical service company, and the Barcelona Supercomputing Center (BSC), which operates Marenostrum, one of Europe's most powerful supercomputers. BSC provided hardware as well as its best computing engineers and scientists for the task.

In 2010, and following the success of the Kaleidoscope Project, BSC and Repsol created a joint research center: the Repsol BSC Research Center

* http://www.bsc.es/projects/kaleidoskope_tmp/index.html.

(RBRC). The aim of the center is to tackle geophysical problems, which was the goal of the previous project, and a broad spectrum of high-performance computing (HPC) problems of interest for Repsol. The geophysical and computational developments at the RBRC have resulted in a unique software platform called Barcelona Seismic Imaging Tools (BSIT),* a whole suite of imaging applications, which includes state-of-the-art solutions for the most challenging problems in exploration geophysics today, which is ready to be used in the latest massively parallel computing infrastructures.

22.1.3 Case Study: 3D Full Waveform Inversion with BSIT

The most advanced geophysical imaging tool in production by the oil industry is Reverse-Time Migration (RTM), which delineates geological structures of the subsurface at a great computing cost (Araya-Polo et al. 2010). RTM was the target of the Kaleidoscope project, and its efficient implementation in powerful computing platforms is the key to its success (Figure 22.1). Full Waveform Inversion (FWI), on top of delineating structures, gives quantitative insight on the properties of the rocks in the subsurface, thus adding a whole new dimension to the insight that can be extracted from the same data (Kormann et al. 2013).

22.1.3.1 Critical Role of the HPC Community

Already known since the 1980s, 3D FWI has not been adopted by the geophysical exploration industry due to its enormous computing cost. Set as an unconstrained inversion of a Terabyte-scale dataset containing

FIGURE 22.1 Target model (top) and Inverted model (bottom) for a synthetic industrial use case.

* http://www.bsc.es/bsit.

billions of unknowns, 3D FWI has been an impossible dream for the oil industry until very recently.

By enabling 3D FWI technology, the information retrieved from the same data increases dramatically. Just as it happens in the last technology generation with RTM, the new imaging technology gives a critical advantage to its owners. Considering the cost of running a survey for several days or the even higher cost of drilling, use of HPC has a relatively negligible cost, especially when compared to the benefit that can be obtained from using it. It is no surprise that in recent months companies such as Total, BP, and Eni have invested in their own supercomputers.

22.1.3.2 Overview of Techniques Used

BSC's in-house BSIT code has been used to implement the FWI algorithm relying on different levels of parallelism. At a shot level, an adjoint method is used to obtain model gradients from data misfits using the current physical model. At a higher level, we need to accumulate results from multiple shots covering different subareas of the subsurface. Once global gradients are generated, an optimization algorithm is executed to improve the current model by a single step. On top of that, a loop is executed, which cycles through iteration steps and a frequency sequence, which solves first the longest wavelengths of the model.

As the complexity of the simulations is of order n^4, doubling the frequency content leads to a 16-fold increase in computational cost. Hence, it has been critical to develop at BSC a multiscale/multigrid approach to tackle the problem efficiently. Furthermore, our advances in preconditioner development and shot codification have allowed us to cut costs both in the number of iterations required for convergence and the number of simulated shots during FWI execution. Together with optimization for HPC architectures and accelerators, the BSC-enabled technology allows us to perform a fully 3D FWI involving thousands of shots up to 8 Hz in less than 20 hours using 60 computing nodes. As a comparison, the acquisition of the data in the survey can take several days and cost tens of millions of dollars. Even worse, a single failed drilling effort can cost the company a hundred million dollars. This example demonstrates the benefit of joining state-of-the-art algorithmic research with specialized code engineering to produce real value to an energy company.

22.2 WHAT'S NEXT AND CONCLUSION

The continuous success of the collaboration between Repsol and the BSC is set to continue in the future by further expanding the already impressive HPC-imaging portfolio of the company. Some of the future development lines will tackle the 3D electromagnetic exploration problem, in particular, when carried out together with seismic exploration. It is believed that the joint inversion of both datasets might lead to a massive leap in terms of understanding the properties of reservoirs and hence having crystal-clear pictures of the earth's interior, which would be a major breakthrough in the oil & gas exploration field for the near future.

REFERENCES

Araya-Polo, M., Cabezas, J., Hanzich, M., Pericas, M., Morancho, E., Gelado, I. et al. (2010). Assessing Accelerator-based HPC Reverse Time Migration. *IEEE Trans. Parallel Distrib. Syst.*, 147–162.

Kormann, J., Rodriguez, J. E., Gutierrez, N., de la Puente, J., Hanzich, M., and Cela, J. M. (2013). Using Power-Model Based Preconditioners for 3D Acoustic Full Waveform Inversion. *SEG Technical Program Expanded Abstracts*, Vol 22, issue 1 (pp. 1105–1109). Houston, TX: SEG.

CHAPTER 23

High-Performance Computing Methods at France's HydrOcean

Erwan Jacquin

CONTENTS

23.1 INTRODUCTION

HydrOcean is a consultancy company and CFD solver (a mathematical software engine embedded in an application) license provider specializing in numerical simulation in fluid dynamics. Composed of 25 high-level engineers, HydrOcean provides design support to industry with the use of innovative numerical simulation tools capable of accurately simulating the most simple to the most complex hydrodynamic phenomena. HydrOcean's services enable clients to save time at the design stage, decrease research costs, improve product performance, and reduce design risks.

HydrOcean uses and distributes a wide range of numerical tools for the whole industry, with a dedicated choice of solvers for marine applications. Part of these tools or solvers are developed by HydrOcean in partnership with the Laboratoire d'Hydrodynamique, d'Énergétique et d'Environnement Atmosphérique (LHEEA) of École Centrale de Nantes (UMR ECN/CNRS 6598). HydrOcean and the LHEEA have set up a highly efficient research partnership, which aims at promoting the most innovative research tools developed in their laboratory and also making them accessible to industrial partners in the form of service provision or through user licenses.

The quality of HydrOcean's expertise is based on its solvers. The use of high-performance research solvers specifically provides innovative solutions to customers, requiring huge computing power. On the other hand, and simplifying to the extreme, the accuracy of the solvers is directly related to the calculation cost. These calculation costs can reach up to thousands of processors for several days. Therefore, high-performance computing (HPC) activities at HydrOcean have three objectives:

1. To be able to simulate very large problems (several hundred million points)
2. To reduce the calculation cost to popularize numerical simulations
3. To decrease the recovery time to make it compatible with industrial processes.

23.2 SPH-FLOW DESCRIPTION

Among the solvers developed by HydrOcean, SPH-Flow solver is one the most advanced in terms of HPC requirements. SPH-Flow implements the smoothed particle hydrodynamics (SPH) method. It is developed 11 years ago at LHEEA and adopted by HydrOcean for the last 7 years. SPH uses a particle method based on a Lagrangian approach without constant connectivity between the elements. This approach is well suited for catching the free surface without any specific extra work. That is why the method is well suited for dynamic flow, with complex and/or deformable bodies, or with highly deformable interfaces that might involve free-surface fragmentation and reconnections. This method aims at being a complement to well-established mesh-based methods and actual commercial solvers (Fluent, CFX, StarCCM+, etc.). The scientific and industrial world is also very interested in the SPH method gathered in the SPHERIC (an international organization representing the community of researchers and industrial users of SPH) group of ERCOFTAC (European Research Community on Flow, Turbulence and Combustion).

23.3 HPC OVERVIEW

SPH methods suffer from high-computational costs that increase dramatically in 3D engineering applications. To increase the industrial use of SPH methods, large simulations must be solved in reasonable return times. Simulations involving several million particles require important computational resources and efficient parallelization.

With the help of the European FP7 NextMuSE project, addressing large parallel SPH simulations and interactive-enhanced visualization, the parallel efficiency of SPH-Flow has been improved to reach high performances on simulations involving up to 3 billion particles and running on 32,768 cores. The parallelization strategy adopted takes advantage of a dynamic domain decomposition in which each subdomain with its underlying particles is assigned to a processor. Interactions between processors are performed using non-blocking MPI (node-to-node Message Passing Interface) communications. During the NextMuSE project, the domain decomposition has been enhanced adopting an ORB (Orthogonal Recursive Bisection) technique, and the MPI communication idle times have been fully masked by optimizing the computation overlapping. The ORB algorithm now enables the decomposition of billions of particles on thousands of cores in only a few minutes, and recent performance tests remove previous parallelization bottlenecks. The final scalability study performed on Switzerland's ETH Zurich machine "Monte Rosa" led to a parallel efficiency larger than 90% and quasi-linear speedups on up to 32,768 cores (Figure 23.1).

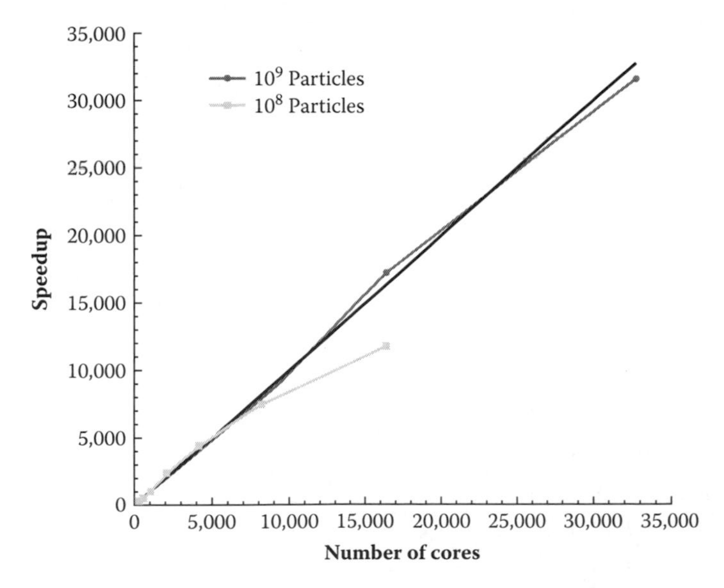

FIGURE 23.1 Speedups for 10 time-steps and one domain decomposition on a 3D dam break problem.

Recently, thanks to the support of GENCI through the HPC for SMB project (in partnership with Inria and BPI France), HydrOcean has improved the code to take advantage of the new many-core architectures such as Nvidia's GPGPU and Intel's Xeon PHI). To optimize the efficiency on these architectures, the algorithms have to be vectorized in order to solve equations simultaneously instead of sequentially. These efforts have been awarded at the supercomputing community's SC13 conference in Denver with the IDC "HPC Innovation Excellence Award" for the high level of innovation reached.

To help develop its massively parallel capabilities, the PRACE European HPC research infrastructure granted nearly 13 million core hours for two research projects tackling complex simulations in the field of sea survivability. HydrOcean was one of the first SMEs (small- and medium-sized enterprises) to benefit from the newly created PRACE OpenR&D offer to industry and thanks to this allocation we jumped from few time-step simulations—used just for performance measurement purposes—to actual cases with long physical durations and hundreds of millions of particles.

Figure 23.2 shows the result of a simulation of a free-fall lifeboat into still water. A correct assessment of loads on free-fall lifeboats during launch is essential for both structural and human safety. When a lifeboat penetrates the sea-surface high-dynamic slamming pressures occur on the lifeboat's hull resulting in a strong (negative) acceleration of the lifeboat. A simulation of this phenomenon was performed using three mesh sizes: 1, 10, and 100 millions of particles. Larger meshes provide more realistic observations of the water jet flow created by the lifeboat free fall. A good capture of the flow requires a very large number of points. This simulation ran on more than 4,096 cores on the "Hermit" CRAY XE6 machine of the HLRS center of Stuttgart. The use of a variable space resolution makes this case of particular interest as it enables to focus on the accuracy in the impact area, and therefore capturing almost otherwise unreachable physical phenomena.

23.4 FUTURE PLANS AND CONCLUSIONS

SPH-Flow solver is very efficient on modern supercomputing platforms, but must be prepared to tackle the petascale challenge. Two axes are currently explored: (1) efficiency of the code on CPUs, so performance can take advantage of peak processor FLOPS and (2) offloading computation to coprocessors like Xeon PHI or GPGPU that are very well adapted to the SPH numerical model.

FIGURE 23.2 **(See color insert.)** Lifeboat impact (1/10/100 millions of particles) application example of SPH-Flow.

HPC is inseparable from the work performed at HydrOcean. It is a key tool for development of HydrOcean and its solvers because it allows providing innovative solvers with acceptable CPU time for industrial clients. Therefore, HydrOcean has set up partnerships with HPC institutes like GENCI, CSCS, IFREMER, and CRIHAN and obtained the support of PRACE. Thanks to our partners, the work done on HPC has achieved international visibility, with, for example, the IDC award obtained in 2013. And while HPC is not the only way to provide CPU resources it is for HydrOcean the way to develop differentiation, to provide innovation to its clients, and then to promote its international development.

III

Industry Developments and Trends

CHAPTER 24

A View from International Data Corporation (IDC)

HPC and HPDA

Steve Conway and Chirag Dekate

CONTENTS

24.1 INTRODUCTION

The first part of this chapter summarizes the historical growth of the worldwide high-performance computing (HPC) market and describes important differences between HPC and commercial business computing. The remainder of the chapter zeroes in on the formative market for high-performance data analysis (HPDA)—big data using HPC—and how this market is bringing the worlds of HPC and commercial business computing closer together. To illustrate the value HPC and its close relative, HPDA, can bring to an organization, the chapter includes first-hand perspectives from two very different adopters who graciously agreed to contribute to the International Data Corporation (IDC) report on which much of this chapter is based:

1. The first perspective is from the National Oceanic and Atmospheric Administration (NOAA), a government organization that, for many years, has relied on HPC to produce its National Weather Service forecasts and other leading-edge work.
2. The second perspective comes from PayPal, a successful, global e-commerce company that adopted HPC not long ago for real-time detection of online fraud and now plans to extend HPC use to affinity marketing and other applications.

24.2 IDC OPINION

HPC, once a niche market serving government- and university-based researchers, began penetrating Tier-1 commercial firms in the late 1970s. HPC quickly established itself as a game changer for accelerating innovation and competitiveness in public- and private-sector organizations. The arrival of commercial-grade clusters in 2001–2002 made HPC affordable even for most small- and medium-size enterprises (SMEs) and startups, with HPC system prices now starting at under $10,000. During the past two decades, HPC has been one of the fastest growing markets in information technology (IT), expanding from $2 billion in 1990 to $21.9 billion in 2012. Few people would have imagined at the dawn of the supercomputer era that HPC systems would be used to help design products ranging from cars and airplanes to golf clubs, potato chips, and diapers—much less to enable a company such as PayPal to detect online consumer fraud in near real time.

IDC forecasts that

- Commercial HPC adoption will continue to ramp up, helping to propel the HPC market to $29 billion in 2017.
- As more companies of all sizes in more markets learn to exploit HPC to speed and improve innovation, competitors lacking this advantage will fall behind.
- Successful corporate chief information officers (CIOs) will need to gain a basic understanding of HPC and ensure that their organizations carefully consider whether to adopt this technology.

24.3 SITUATION OVERVIEW

IDC offers 10 essential considerations in the following sections to describe the contemporary realities and the potential value of HPC, especially for CIOs.

24.3.1 HPC: 10 Essential Considerations

24.3.1.1 HPC Is One of the Fastest-Growing IT Markets

Between 1990 and 2012, revenue for the worldwide HPC ecosystem—servers, storage, software, and services—ballooned more than tenfold, from $2 billion to $21.9 billion (see Figure 24.1).

Propelled by the standard-based cluster—an HPC innovation from America's NASA—the HPC server market expanded during the decade of the 2000s faster than the "hot" IT markets for flat-panel TVs or online gaming. In addition

1. In 2009, the worst year of the global economic recession, revenue for supercomputers priced at $500,000 and above grew 35% and revenue for high-end supercomputers selling for $3 million and above jumped a whopping 65%.
2. The worldwide HPC server market posted record revenue in 2011 and 2012, with server systems costing $500,000 and up achieving a 29.6% year-over-year revenue gain.

The rapid growth of the HPC market in recent years has attracted new vendors to this market and has caused some of the largest existing IT vendors, including HP, IBM, Intel, EMC, Dell, Bull, and others, to ramp up their HPC activities.

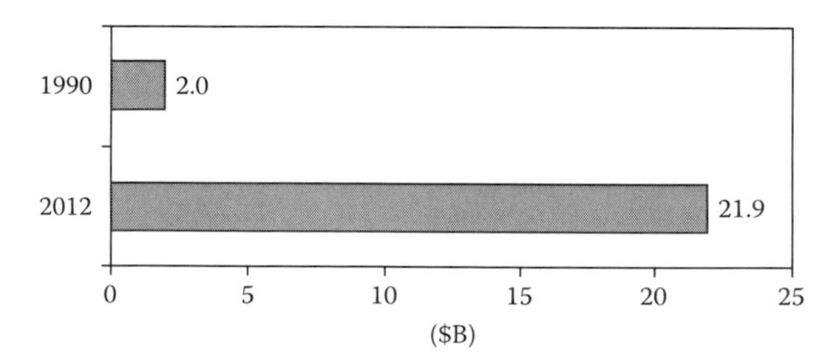

FIGURE 24.1 Worldwide HPC revenue, 1990 & 2012. (From IDC, 2013.)

24.3.1.2 Commercial Firms Began Adopting HPC in the 1970s

Following its initial growth spurt in the late 1960s and 1970s, the market for HPC systems has expanded over time by adapting to the requirements of successive waves of new users—in large part through advances in software. Each new wave of users has expected HPC system vendors to do more for them, by providing software to make these systems easier to deploy and use.

In 1976, the CRAY-1 supercomputer was delivered to its first customer, Los Alamos National Laboratory (LANL), as a blazingly fast hardware platform with no operating system. Not to worry: LANL and others in the first wave of HPC users, primarily government and university researchers, typically had enough in-house technical savvy and personnel to write software themselves when the need was critical.

The second wave of adoption carried HPC into industry, initially the automotive, aerospace, and petroleum sectors, starting in the late 1970s. These users required HPC vendors not only to provide an operating system and other system software, but also to port the key third-party ISV applications needed to run the users' industry-specific problems—and to run the applications with the reliability expected in production computing environments.

The third adoption wave, driven by the compelling price/performance of standard-based clusters, began in 2001–2002 and continues today. This phase greatly expanded the market for HPC by making this game-changing technology affordable and tractable for less-experienced users, such as financial services firms, consumer product makers, and online companies of nearly any size. Figure 24.2 illustrates the range of companies using HPC today.

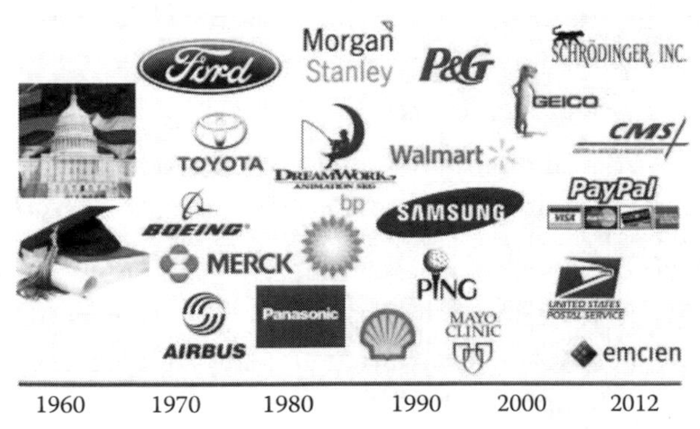

An illustrative HPC adoption history, with commercial HPDA users to the right.

FIGURE 24.2 HPC adoption timeline. (From IDC, 2013.)

24.3.1.3 Ninety-Seven Percent of Adopters Say HPC Is Indispensable for Their Ability to Compete and Survive

In a worldwide IDC study conducted for the Council on Competitiveness, based in Washington, DC, 97% of the commercial firms that had adopted HPC said they could no longer compete or survive without it. The chief benefit cited by these firms is that HPC enables them to bring more innovative, higher-quality products and services to the market in shorter time frames. Whether the task is to design a commercial airplane, discover an oil reservoir in deep water, optimize a portfolio of mortgage-backed securities, create a feature-length animated film, develop a titanium golf club, or detect patterns of online consumer fraud, HPC allows more possibilities to be explored within the given time frame. In many cases, HPC makes it possible to solve problems that could not be solved otherwise. As the Council on Competitiveness puts it, “to out-compute is to out-compete.”

24.3.1.4 Senior Government Officials Increasingly Recognize HPC’s Economic Value

HPC is important for national economies, because HPC also called supercomputing, has been firmly linked to economic competitiveness as well as scientific advances. As noted previously, 97% of companies that had adopted HPC told IDC they could no longer compete or survive without it. Worldwide political leaders increasingly recognize this trend:

- In his 2006 State of the Union address, U.S. President George W. Bush promised to trim the federal budget, yet urged more money for supercomputing.
- In 2009, Russian President Dmitry Medvedev warned that without more investment in supercomputer technology, Russian products “will not be competitive or of interest to potential buyers.”
- In June 2010, Representative Chung Doo-un of South Korea echoed that warning: “If Korea is to survive in this increasingly competitive world, it must not neglect nurturing the supercomputer industry, which has emerged as a new growth driver in advanced countries.” The Korean National Assembly then called for the creation of a national 5-year plan for advancing HPC.

- In his 2011 State of the Union address, President Obama noted China's rapid progress in HPC and said that the U.S. Department of Energy's Oak Ridge National Laboratory is "using supercomputers to get a lot more power out of our nuclear facilities."
- In February 2012, the European Commission announced that it had adopted a plan to double spending on HPC to €1.2 billion, with much of that money aimed at the installation of additional large supercomputers at leading European HPC centers. In 2011, at the European Commission's request, IDC had recommended a 5-year HPC strategy for Europe to implement.

24.3.1.5 HPC Is Different from Business Computing

IDC uses the terms *high-performance computing* and *technical computing* as synonyms to encompass the entire market for computer servers (and related software and services) used by scientists, researchers, design engineers, analysts, and others to address computationally intensive or data-intensive modeling, simulation, and other analytical problems. HPC activities can be found in the commercial sector, government, and academia. Commercial activities include automotive and aerospace product development, oil and gas exploration, drug discovery, weather prediction and climate modeling, complex financial modeling, consumer product design and optimization, and advanced 3D animation, as well as Big Data analytical problems in many commercial domains.

In contrast to HPC, commercial computing is used for business operations such as accounting, payroll, sales, customer relationship management (CRM), enterprise resource planning (ERP), transaction processing, human resources, and purchasing. The common denominator underlying HPC problems is a degree of algorithmic complexity that is atypical for business IT problems. Business IT workloads often (though not always) consist of a large volume of tiny problems—for example, a business computer may process tens of thousands of transactions per second. In contrast, a single HPC problem may take not a fraction of 1 second but hours, days, weeks, or even months to process. HPC systems are designed to support very large, long-running, input/output (I/O) intensive problems.

24.3.1.6 Goals of IT and HPC Are Also Different

CIOs who are new to HPC often make the mistake of treating it like a typical IT function. This misperception can lead to lost productivity and, in some cases, to conflict between the CIO's office and the company's HPC staff.

As Jim Barrese, CTO of PayPal, advises in his commentary later on in this chapter:

> "Clearly understand that HPC is not a mass consumption technology where we enable everyone in our organization with it. This is a deep engineering function. It's custom built and includes writing software to solve cutting-edge problems... Think of HPC not as an IT function but as a competitive business advantage. There's a hard link between HPC and PayPal's top line and bottom line."

In important respects, HPC is different from general IT deployments. IT is generally about *provisioning*—equipping each of the company's knowledge workers with the basic computing tools they need to perform their jobs productively, and providing as little beyond that as possible to stay within the budget. HPC, on the other hand, is about *enablement*—providing a small subset of specialized knowledge workers with the most powerful computational tools the company can afford. A typical IT worker's desktop or laptop system is capable of fully supporting the worker's computing requirements, while there is often no limit to the amount of computing power an HPC user could exploit on the company's behalf.

Reflecting the almost insatiable demand for HPC resources, utilization for most HPC systems exceeds 90%, compared with about 30% for typical business servers. The powerful trend toward server consolidation and virtualization in enterprise datacenters has had almost no impact in HPC datacenters—because there are few unused cycles available to consolidate and exploit through virtualization.

24.3.1.7 Key IT Datacenter Technologies Have Trickled Down from HPC

There is a perennial debate between those who argue that key IT technologies "bubble up" from the low end, such as embedded and desktop devices, and those who counter that key technologies "trickle down" from the high end, especially HPC. In reality, of course, both arguments are correct. Technological innovation is bidirectional, flowing up and down. In addition

- During the decade from 2000 to 2010, for example, clusters based on standard ×86 processors from Intel and AMD supplanted RISC processor-based computers to become the dominant species of HPC systems. The ×86 processors bubbled up from the market for desktop/laptop computers.
- Conversely, clusters themselves were born in the HPC market and later trickled down into enterprise IT datacenters. The original Beowulf cluster was developed at NASA in 1994 by Thomas Sterling and Donald Becker, although standard-based clusters did not begin to gain strong market traction in HPC until 2001–2002.
- The Linux operating system has played a major role in making clusters dominant in HPC. Soon after their adoption by leading-edge HPC sites, Linux clusters began moving into commercial datacenters. Among the first commercial adopters were the back offices of investment banks and other large financial service firms, where "quants" used the Linux clusters for complex tasks including global risk management, pricing exotic instruments, and optimizing investment portfolios.
- Grid computing and cloud computing are two more important technologies that have trickled down from HPC to mainstream commercial markets. And today, as the PayPal extract in Section 24.4.2 attests, commercial companies are increasingly using HPC systems and approaches to tackle daunting problems such as real-time detection of online consumer fraud in high-volume Big Data environments.
- On the bubble-up front, multiple processors and coprocessors have been making their way from the embedded systems market into HPC, including GPUs, ARM, and Atom devices.

24.3.1.8 HPC Systems Now Start at Under $10,000

At one time, decades ago, entry pricing for a supercomputer was in the $25 million to $30 million range. Many people think buying an HPC system still means ponying up millions of dollars. But thanks to the transition to clusters based on industry-standard technologies, pricing for HPC systems now starts at less than $10,000. In addition

- In 2012, 104,148 systems—78% of all HPC systems shipped—were sold for under $100,000. The average price paid for one of these workgroup HPC systems was $15,375.

With entry prices this low, HPC systems have become affordable for many more companies than ever before.

24.3.1.9 Commercial Firms Are Also Adopting HPC for Challenging Big Data Problems

High-performance data analysis is the term IDC coined to describe the convergence of the established data-intensive HPC market and the high-end commercial analytics market that is starting to move up to HPC resources. In addition

- A good commercial example is PayPal, a multibillion-dollar eBay company, which not long ago integrated HPC servers and storage into its datacenter workflow to perform sophisticated fraud detection on eBay and Skype transactions in real time. Real-time detection can catch fraud before it hits credit cards. IDC estimates that using HPC has allowed PayPal to save hundreds of millions of dollars to date.
- Another commercial adopter is GEICO, which is using HPC to perform weekly updates of insurance quotes for eligible U.S. households and individuals.

Simulation-driven HPC is the longest standing part of the HPDA market. Since the start of the supercomputer era in the 1960s, important HPC workloads, such as cryptography, weather, and climate research, have been data-intensive. The newer kid on the HPDA block is analytics, which comes in many flavors. Of course, the financial industry has been running analytics on HPC systems at least since the late 1980s. But newer methods, from MapReduce/Hadoop to graph analytics, have greatly expanded the opportunities for HPC-based analytics.

The common denominator underlying simulation- and analytics-based HPDA workloads is a degree of algorithmic complexity that is atypical for transaction processing-based business computing. With the help of sophisticated algorithms, HPC resources are already enabling established HPC users, as well as commercial adopters such as PayPal, to move beyond "needle in a haystack" searches to discover high-value, dynamic patterns.

IDC believes that HPC resources will be increasingly crucial for extending Big Data capabilities from *search* to *discovery*.

IDC forecasts that revenue for HPC servers acquired primarily for HPDA use will grow robustly, increasing from $739 million in 2012 to exceed $1.4 billion in 2017. Revenue for the whole HPDA ecosystem, including servers, storage and interconnects, software, and service should double the server figure alone. A good chunk of this revenue will come from commercial firms.

24.3.1.10 There Is More on Tap from HPC

One of the next important developments IDC expects to come out of the HPC market is more capable network technologies to speed communications between cores, processors, servers, and nodes. This development should help to address the so-called memory wall, the growing gap between escalating processor peak speeds and the lagging ability of internal networks to feed processors with enough data to keep them busy. Improving network bandwidths and latencies should be especially important for challenging Big Data tasks faced by businesses and government organizations alike.

Think here not only of switch vendors such as Mellanox and Cisco but also of initiatives by processor vendors to move up a level of integration to provide capable fabrics. AMD's SeaMicro initiative comes to mind, along with Intel's acquisitions of QLogic's InfiniBand assets and Cray's proprietary interconnect assets. IDC expects significant progress to occur on this front in the next 5 to 6 years, with benefits for both enterprise and HPC datacenters.

24.4 CASE STUDIES

24.4.1 The National Oceanic and Atmospheric Administration: HPC as a Mission-Critical Tool

Excerpts from an interview with David Michaud, deputy director, High-Performance Computing and Communications (HPCC), NOAA:

> The National Oceanic and Atmospheric Administration's mission is to understand and predict changes in the Earth's environment, from the depths of the ocean to the surface of the sun, and to conserve and manage coastal and marine resources. From daily weather forecasts, severe storm warnings, and climate monitoring to fisheries management, coastal restoration, and supporting marine commerce,

> NOAA's products and services support economic vitality and affect more than one-third of America's gross domestic product. HPC has long been indispensable for carrying out NOAA's mission.

At NOAA, HPC is well integrated with the leadership and management of the typical enterprise IT services. Our CIO has two titles, CIO and director of high-performance computing and communications. So these subtleties are dealt with directly under a consistent perspective.

HPC is a critical tool needed for NOAA to carry out its mission on a daily basis. The extent of NOAA's HPC capability and capacity is directly linked to our mission performance. To ensure our HPC is acquired, implemented, and allocated for the maximal benefit to our organization, we have developed a well-established governance structure. The CIO and director of HPC chairs our High-Performance Computing Board, which includes senior leaders from all of NOAA's major missions, such as weather services, satellite services, ocean services, fisheries services, and our applied research groups. These representatives who participate in overseeing HPC are generally non-technologists who are responsible for setting programmatic priorities within their mission areas. Beneath this group are the technologists responsible for acquiring HPCC systems and operating them on a daily basis, along with program managers in charge of allocating and prioritizing the HPC resources.

From the CIO's perspective, the highest priority is IT security, but HPC is a close second. Without HPC, we cannot perform our mission work. NOAA's senior leaders recognize that HPC is a key component of our value chain for producing NOAA's weather reports and other products. Weather observations are very important, and HPC is important for turning all that data into meaningful information. We invest over a billion dollars each year in observations. HPC is crucial for leveraging those investments. The final crucial element is dissemination, getting the data out to the public.

At its core, HPC needs to be about enabling an organization's mission, not just putting together a computing program for technology's sake. There are important benefits, even for organizations that are just starting out in HPC. You can use a simple cluster to aggregate disparate applications within an organization and increase the performance of those applications. Within an HPC system, you can also aggregate all the data sources in one place and extract value from them in one place. These activities often reveal or create previously unknown dependencies between applications, which tighten relationships among groups within an organization. All of these benefits yield cost efficiencies. And while you should not expect all the benefits to be realized immediately, even in the first year, you should expect to see important benefits with some targeted applications.

Thanks to HPC's strong connection to our CIO, at NOAA, we are able to provide HPC from an enterprise perspective. We started this integration of HPC into enterprise IT in 2006. Now, NOAA is looking to collaborate across federal agencies by providing HPC resources in a shared services paradigm. We are really good at knowing how to exploit midsize HPC systems cost–effectively in a high reliable manner, and we can help other organizations that have similar needs. One part of this challenge is working with a mix of experienced and less-experienced users. In turn, we leverage other federal agencies' facilities such as the Department of Energy and National Science Foundation to understand how to exploit their larger, leadership-class HPC systems.

24.4.2 PayPal Exploits HPC for Fraud Detection

Excerpts from an interview with James Barrese, CTO, PayPal:

> PayPal is a wholly-owned subsidiary of eBay Inc., a company that in 2012 earned $14 billion in revenue and enabled more than $175 billion in commerce. Among other things, PayPal is responsible for detecting fraud across eBay and StubHub. By deploying HPC servers, storage, and software, PayPal is able to detect fraud in near real time before it hits consumers' credit cards. HPC has helped PayPal to catch substantial additional fraud.
>
> PayPal's decision to use HPC goes back to our innovative technology roots. More than $5,200 is transacted every second across our platform, and it's our responsibility to not only deliver great customer experiences but keep our customers' financial information safe. As our company continued to rapidly grow, we saw both a challenge and an opportunity with handling fraud and risk while delivering more personalized shopping experiences. Our approach to these types of situations is to be very entrepreneurial and aggressive. We asked, how can we solve this business problem, and our HPC team said, "Hey, this problem's been solved before, using HPC." This led to an HPC pilot and then to deployment.
>
> I think we're in the most interesting time in my career—the pace of technology change is accelerating. There's a whole new S-curve with being able to leverage HPC and machine learning on

a real-time basis. HPC helps us manage large, globally-distributed systems. With HPC, we're able to do risk analysis in real time and detect fraud before it happens. We're also able to leverage HPC to further personalize the consumer experience. Say you walk by a Jamba Juice and we know you like smoothies, so we deliver a coupon to you right then and there.

When you are evaluating whether or not HPC is right for your company, here are some tips:

- You need the right talent to be able to deliver results. I would urge people to start with clear objectives and have the right people on board. We had the nucleus of HPC talent and have been recruiting more HPC talent aggressively.
- Clearly understand that HPC is not a mass consumption technology where we enable everyone in our organization with it. This is a deep engineering function. It is custom built and includes writing software to solve cutting-edge problems.
- Think of HPC not as an IT function but as a competitive business advantage. There is a hard link between HPC and PayPal's top line and bottom line.
- HPC is key to our company's strategy because PayPal is redefining an industry, and it is all driven by technology. HPC is enabling new business capabilities that are really product driven.

24.5 FUTURE OUTLOOK

Especially during the 2000–2010 decade, the compelling price/performance of standard-based clusters made them the dominant species of HPC systems and greatly expanded the size of the HPC market. For the first time, even SMEs and start-ups could afford to move up to HPC. Today, HPC technology is helping companies and government organizations of all sizes to innovate, compete, and survive.

IDC forecasts that the HPC market will continue to grow faster than the market for general business servers, owing partly to the irrelevance of server consolidation in HPC. Existing commercial users of HPC will expand their use, and more companies will adopt HPC for the first time to tackle daunting business challenges. On the business analytics/business intelligence side, IDC expects more firms to follow the examples of PayPal and GEICO in deploying HPC for high-value problems that cannot be effectively addressed with enterprise server technologies alone.

IDC believes that CIOs will increasingly be expected to become familiar with the benefits of HPC, whether they decide to adopt this technology or not. Because of the proven game-changing value of HPC, CIOs will also increasingly be expected to know which of their competitors is exploiting HPC.

24.6 ESSENTIAL GUIDANCE

- CIOs should gain a basic understanding of HPC and ensure that their organizations carefully consider whether to adopt this technology. CIOs who do not assess the potential benefits of HPC will increasingly risk losing ground to competitors who have learned how to exploit HPC to accelerate and improve innovation.
- HPC is no longer an ultraexpensive, arcane technology. Prices for HPC systems now begin at less than $10,000, and the vast majority of HPC systems are sold for less than $100,000—although the biggest supercomputers fetch more than $100 million each. HPC systems are also much easier to install, deploy, and manage than in former times. They ceased long ago to be the exclusive province of white-coated lab scientists.
- HPC should not be treated in the same way as other corporate IT functions. If general IT provisioning is a gunshot, then HPC enablement is a more highly targeted rifle shot. HPC has a much narrower focus than general IT and is typically meant for a small subset of an organization's knowledge workers. But as IDC studies consistently confirm, HPC can have a much greater strategic impact on organizations that exploit this game-changing technology.

24.7 HPDA: BIG DATA MEETS HPC

The benefits of HPC increasingly depend on the ability to analyze large, heterogeneous volumes of data in brief periods. The nature of scientific activity is rapidly changing. Instead of lone individuals working in isolation, teams of scientists now collaborate across national and other geographic boundaries in pursuit of advances. Providing distributed access to data from powerful scientific instruments and sensor networks, along with access to powerful HPC systems, is critical for this mode of collaboration. Both the scientific instruments and the HPC systems themselves generate data volumes that are unprecedented in size and variety. The split-second need to interpret sensor data to prevent a local power grid failure from cascading illustrates the time-criticality of certain HPC big data problems.

Industry is also in the midst of a twenty-first-century revolution driven by the application of computer technology to industrial and business problems. Given their broad and expanding range of high-value economic activities, HPC users are increasingly crucial for industrial innovation, productivity, and competitiveness. HPC users pursue industrial product design with *virtual prototyping and large-scale data modeling* (i.e., using computers to create digital models of products or processes and then evaluating and improving the design of the products or processes by manipulating these computer models). HPC users may also employ analytics methods to help uncover meaning in the data itself.

The growing market for HPDA—using HPC for data-intensive challenges—is already enlarging HPC's contributions to science, commerce, and society. HPDA promises to play a major role in helping to address the major opportunities and challenges of the twenty-first century.

24.8 THE ORIGINS OF DATA-INTENSIVE COMPUTING

Handling large data volumes with calculating machines is not new. A noted antecedent was Herman Hollerith's tabulating machine, which was used to process the 1890 U.S. census data from punch cards. The Hollerith cards look remarkably like the IBM punch cards that became a standard input medium for machine data after their introduction in 1928. But for nearly a century before Hollerith adopted this idea, punch cards had been used to automate the operations of the Jacquard loom (1801) and derivative textile machines. The notion of a programmable calculating machine for processing daunting data harks back at least as far as the unfinished difference engine of Charles Babbage (1791–1871). A meandering historical line leads from the Jacquard loom, Babbage's difference engine, and the Hollerith tabulator to the dawn of the supercomputer era in the 1960s–1970s, when punch cards were still being used to feed data into the CDC 6400, CDC 6600, and optionally the Cray-1 computer (which could also ingest data from digital magnetic tape).

For decades, the HPC community has loosely divided workloads into two categories. *Compute-intensive* jobs call for substantial computing to be performed on relatively small data volumes. *Data-intensive* jobs reverse the ratio of computing to data such that the data volumes predominate.

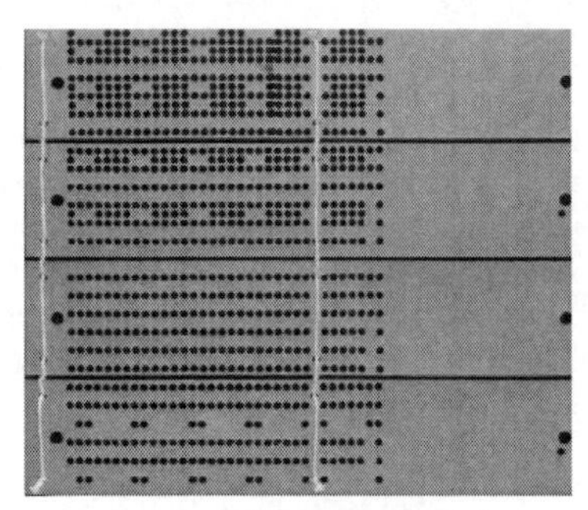

Jacquard loom punch card

The reading board for a Hollerith punch card (1890)

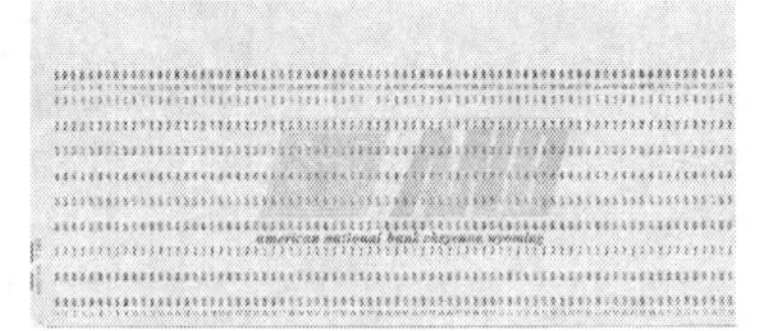

Typical 80-column IBM punch card (circa 1970)

The historical evolution of data-intensive computing received a powerful thrust from the military and defense needs of World War II (1939–1945). Necessity proved to be the mother of invention as data-intensive applications crucial for the war efforts of the major combatants—especially cryptography, ballistics, and weather forecasting—produced technical innovations that laid the groundwork for the modern computer industry. The military and defense needs of the Cold War (1945–1991) continued the government funding streams that played a primary role in driving advances in scientific-technical computing, eventually spawning the supercomputing (HPC) industry in the 1960s. It is no coincidence that many computer industry pioneers, including John von Neumann, Alan Turing, and Seymour Cray, were funded to contribute to their countries' military and defense initiatives.

24.8.1 Data-Intensive Computing Categories

In the supercomputer era, most data-intensive computing jobs in the government, academic, and industrial sectors have involved modeling and simulation—that is, using digital representations to analyze, and in some cases manipulate and optimize, the behavior of complex physical and quasi-physical systems. The systems in question range from product designs for cars, planes, golf clubs, and pharmaceuticals, to subatomic particles, global weather, and climate patterns, and the cosmos itself.

But from the start of the supercomputer era, an important subset of HPC jobs has involved analytics—attempts to uncover useful information and patterns in the data itself. (Prescriptive grammarians: we follow the *Wall Street Journal*'s example by treating "data" as a collective, singular noun.) Cryptography, one of the original scientific-technical applications, falls predominantly into this category.

HPC-based analytics has not been limited to government and academia, however. The financial services industry was the first commercial market to adopt supercomputers for advanced data analytics. In the 1980s, large investment banks began hiring particle physicists from Los Alamos National Laboratory and the Santa Fe Institute to employ HPC systems for daunting analytics tasks, such as optimizing portfolios of mortgage-backed securities, pricing exotic financial instruments, and managing firm-wide, global risk. This practice has continued: in 2013, Goldman Sachs lured a particle physicist away from the Large Hadron Collider work at CERN. High-frequency trading is a newer addition to the financial services industry's library of HPC-enabled applications.

24.8.2 High-Performance Data Analysis

High-performance data analysis is the term IDC coined several years ago to embrace all data-intensive computing that requires HPC resources, including workloads based on modeling and simulation, and workloads involving analytics. The term is a research-tested compromise that recognizes the preferences of the HPC community, for whom analytics is only one form of analysis, and the preferences of a new wave of commercial firms that are turning to HPC for the first time to tackle daunting business analytics and business intelligence problems. The commercial buyers and their vendors typically like the term *high-performance analytics* but few are ready yet to categorize what they are doing as *high-performance computing*. For at least the near term, HPDA can serve as a rubric to describe the convergence of the established data-intensive HPC market based on modeling and simulation, and the advanced analytics market that increasingly includes commercial firms moving up to HPC resources for the first time. In a nutshell, think of HPDA as meaning big data problems that need HPC.

24.8.3 What Is Driving Demand for HPDA?

HPDA is an evolutionary and a revolutionary story. The data explosion fueling the growth of HPDA stems from a mix of long-standing and newer factors:

- The ability of increasingly powerful HPC systems to run data-intensive modeling and simulation (M&S) problems at larger scale, at higher resolution, and with more elements (e.g., inclusion of the carbon cycle in climate ensemble models).
- The proliferation of larger, more complex scientific instruments and sensor networks, from "smart" power grids to the Large Hadron Collider and Square Kilometer Array.
- The increasing transformation of certain disciplines into data-driven sciences. Biology is a notable example, but this transformation extends even to humanities disciplines such as archeology and linguistics.
- The growth of stochastic modeling (financial services), parametric modeling (manufacturing) and other iterative problem-solving methods, whose cumulative results produce large data volumes.
- The availability of newer advanced analytics methods and tools: MapReduce/Hadoop, graph analytics, semantic analysis, knowledge discovery algorithms, and others.
- The escalating need to perform advanced analytics in near real time—a need that is causing a new wave of commercial firms to adopt HPC for the first time.

24.8.4 Running Modeling and Simulation Problems with Greater Realism

Contemporary HPC systems have made it feasible to tackle data-intensive M&S problems at scales and resolutions that could only be imagined before, and with more elements included. Rapidly expanding HPC computational and storage sizes have triggered explosive growth in the data ingestion, data processing, and data outputs of high-performance computers. It is usual to encounter HPC data centers with multi-petabyte disk storage capacities and even larger tape facilities.

A leading example is the "BlueWaters" supercomputer at the National Center for Supercomputing Applications (NCSA) (University of Ilinois), which features 25 petabytes of disk storage and 300 usable petabytes of tape capacity. True, storage access speeds have not nearly kept pace with capacity advances, but the overall trend has placed increasing emphasis on the data side of HPC. Many scientific and engineering M&S problems have been able to benefit from recent advances in system capabilities. But a 2012 IDC worldwide study for the National Science Foundation (United States) and the NCSA documented the frequent desire for even greater M&S realism by HPC users in industry, government, and academia—along with the obstacles to increased realism.

Helping to drive the need for more capable HPC systems is the proliferation of more powerful scientific instruments and sensor networks. Extreme cases include the Large Hadron Collider at CERN, which generates one petabyte of data per second when running, and the planned Square Kilometre Array telescope, which will produce one exabyte of data per day when it becomes fully operational in 2024. (In both cases, only a small fraction of the raw data output is slated to be stored.) More mainstream examples are initiatives around the world to employ HPC for real-time control of smart power grids, including efficient power distribution, failure prevention and recovery, and other crucial requirements.

24.8.5 Expanding Analytics Use in Established HPC Fields

Newer methods, especially graph analytics, semantic analysis, and knowledge discovery algorithms, as well as tools such as MapReduce/Hadoop, are quickly expanding the role of analytics in established HPC disciplines and markets. Some of the longest-standing HPC domains, including climate science, bioscience, and energy research, are actively assessing the efficacy of these analytics approaches.

In some cases, existing HPC users seek to maximize insights and innovation by applying both established M&S and newer analytics methods to the same problem, often using the same HPC system. A prominent example is climate research, one of the most data intensive of all HPC domains. The report from the first Climate Knowledge Discovery Workshop (April 2011, Hamburg, Germany), nicely sums up this trend:

> Current approaches to data volumes are primarily focused on traditional methods, best-suited for large-scale phenomena and coarse-resolution data sets. The data volumes from climate modeling will increase dramatically due to both increasing resolution and number of processes described. What is needed is a suite of new techniques

> [i.e., knowledge discovery algorithms] interpreting and linking phenomena on and between different time- and length scales as well as realms and processes. Such tools could provide unique insights into challenging features of the Earth system, including extreme events, nonlinear dynamics, and chaotic regimes.

The climate research community's vision of extracting new insights by applying existing and new algorithm types to their problems—for the most part separately, not as a mashup—points the way toward what more of the established HPC domains will do.

Perhaps no field has stronger potential for benefiting from HPC-based analytics than bioscience. Data-intensive applications already in motion in this varied field range from advanced research—notably in genomics, proteomics, epidemiology, and systems biology—to commercial initiatives to develop new drugs and medical treatments, agricultural pesticides, and other bioproducts.

One of the world's most socially and economically important HPDA thrusts is certainly the multiyear transition from today's procedure-based medicine to personalized, outcome-based health care. Identifying highly effective treatments in near real time by comparing an individual's genetic makeup, health history, and symptomology against tens of millions of archived patient records poses enormous HPDA challenges that may take another decade to master. When this capability matures, it will likely serve as a decision-support tool of unprecedented utility for the global health-care community.

The bioscience discussion provides a good excuse to remind ourselves that progress in science, including HPDA-enabled science, may use shortcuts. In 2003, for example, the sequencing of the human genome (following a draft version in 2000) was rightly hailed as a breakthrough that would launch a new era in bioscience research. But the nominally complete version, it turned out, targeted only the euchromatic regions of the genome deemed capable of coding—transcribing DNA to RNA and to the proteins needed for our biological functions. The weaker-staining, heterochromatic regions were harder to characterize with available techniques and were ignored as so much evolutionary space junk.

Believing that evolution must be more economical than this, a few years later, researchers tackled the heterochromatic regions and effectively completed the sequencing of the full human genome. Although the precise functions of these areas are not settled yet, evidence strongly suggests that the heterochromatic regions play important roles in gene expression and in maintaining order within the human genome's 3 billion base pairs of chromosomes. Not space junk after all.

This episode illustrates how science sometimes advances in half steps, temporarily simplifying hairy problems for the sake of expediency before returning better armed to attack them head on. In the extreme case of systems biology, the known science is so rudimentary compared to the fathomless complexity of the subject matter that much of the research activity qualifies only as prescience—because the possibilities are still expanding, rather than contracting as required in scientific pursuits.

Climate research and bioscience are not the only HPC vertical market segments that will benefit from advances in analytics. In truth, newer analytics methods and tools are likely to benefit all existing HPC vertical segments at least to some extent. These segments also include computer-aided engineering, chemical engineering, digital content creation and distribution, electronic data automation, financial services, geosciences and geo-engineering (oil and gas), defense, government labs, and academia. But the story does not end there.

24.8.6 Fraud Detection, Cyber Security, and Insider Threats

High-potential horizontal analytics applications are also starting to make an important impact in the world of HPC. Fraud detection, cyber security, and insider threats are increasingly crucial challenges for established HPC users in government, academia, and industry to meet—and they are causing a new wave of commercial organizations to move up to HPC for the first time.

Historically, a major challenge of civilizations has been to safeguard items of high value from external plunder and insider theft. Most ancient cities were fortified sites built to protect stores of surplus food and other raw materials on which the lives of urban populations depended. The stored wealth of cities has always been subject to brute force attacks by hostile armies, Trojan horse-style intrusions, and betrayals to enemies by treacherous city residents.

These same threats have now been extended to a newer type of stored value: the vast volumes of data that describe and help to maintain critical infrastructures—national security systems, power and communications systems—along with private data on large government programs (e.g., social security, Medicare, and Medicaid) and data containing industrial trade secrets. In this Big Data context, the main threats fall into the following categories:

- Cyber-attack: An attempt by an outside party to gain unauthorized access to stored data or disrupt services, usually via network intrusion (hacking). Cyber-attackers have breached the U.S. electrical grid, planting potentially harmful software in the process. In March 2013, attackers hijacked the servers of Dutch web-hosting firm Spamhaus to carry out the largest cyber-attack in history—an attack so powerful and widespread that "it almost broke the Internet," according to the firm hired to subdue the attack. The head of the Pentagon's Cyber Command, Gen. Keith Alexander, estimates that cyber-attacks and intellectual property theft cost U.S. companies $250 billion annually.
- Insider threat: The unauthorized appropriation or exploitation of data by a person with access to the organization owning the data. Espionage is one insider threat category in government, as this example from an FBI brochure confirms: "Greg Chung spied for China from 1979-2006. Chung stole trade secrets about the space shuttle, the Delta IV rocket and the C-17 military cargo jet for the benefit of the Chinese government . . . In February 2010 he was sentenced to over 15 years in prison." Insider theft is also a serious problem in the private sector. In one of many examples, in 2011, hedge fund founder Raj Rajaratnam was sentenced to 11 years in prison for insider trading that netted him $60 million.
- Fraud: The deceptive exploitation or annotation of data for wrongful or illegal personal gain. Example: Estimates for annual losses to fraud by U.S. businesses run as high as $1 trillion. The FBI estimates that 10% of transactions in federal health care programs—Medicare, Medicaid, Veterans Affairs, and so forth—are fraudulent, costing more than $150 billion per year. The banking and financial services industry accounts for 16% of all fraud cases in the private sector, more than any other industry, according to a recent report from the Association of Certified Fraud Examiners.

Tackling these problems requires moving beyond today's needle-in-a-haystack, static searches for items already known to exist in a database. The essential challenge presented by these problems is to discover hidden patterns and relationships—things you did not know were there—in real time or near real time, and then to track existing and new patterns dynamically as they form and evolve. This is typically a far more difficult undertaking than static searches, one that traditional technologies such as RDBMS and standard cluster computers (clusters) are not designed to handle well. It entails using graph analysis, sometimes in combination with other methodologies such as semantic or statistical analysis. HPC systems with large shared memories and turbo-charged memory and I/O capabilities have shown a distinct advantage in winning business for this important, emerging class of problems.

24.8.7 A New Wave of Commercial Adopters

A growing number of commercial firms are moving up to HPC for the first time to tackle advanced analytics problems that enterprise computing solutions cannot solve effectively. These problems exhibit algorithmic complexity that is atypical for transaction processing-based business computing. In many cases, the need to solve problems in near real time adds to the appeal of adopting HPC resources and the HPDA methods they enable.

Fraud and error detection performed in near real time, typically using graph analysis, is already becoming an economically important HPDA application and, for vendors, a pursuable HPDA market. The following list of examples hints at the larger convergence of traditional HPC users and new commercial users around a closely related set of HPDA requirements:

- eBay subsidiary PayPal (as described above). IDC estimates that using HPC has saved PayPal more than $700 million to date. The fraud detection initiative has been so successful that PayPal is now in the process of applying HPC to two other major tasks: affinity marketing and managing PayPal's overall IT infrastructure.
- Italy's large government agency, Istituto Nazionale della Previdenza Sociale (INPS), acquired a supercomputer and advanced analytics software to attack health care fraud on a national basis. China's national health care system is following a similar course.
- U.S. health care data is spread across five gigantic databases and the government recovers only about $5 billion per year of the estimated $150 billion per year in fraudulent transactions. The Centers for Medicare and Medicaid Services (CMS) has evaluation projects running at Oak Ridge National Laboratory and the San Diego Supercomputer Center, both aimed at unifying disparate databases for

these federal health care programs and using powerful supercomputers to perform fraud detection across the unified results. This solution could save $50 billion per year by analyzing the data in near real time.

- Every day, the U.S. Postal Service (USPS) delivers up to 4 billion mail scans and delivers mail to as many as 145 million addresses. Several years ago, USPS began buying HPC systems to identify mail with erroneous or fraudulent postage. In 2012, this independent federal agency upgraded to more powerful HPC systems.

24.8.8 Other Commercial HPDA Applications

But fraud and error detection is only one of the high-potential applications in the nascent HPDA market. Consider the following examples:

- Schrödinger is a global life sciences and materials science software company with offices in Munich and Mannheim, Germany. The company also uses molecular dynamics to identify promising candidates for new drugs to combat cancer and other diseases. Schrödinger hired Cycle Computing, a cloud-computing services company, to test 21 million drug candidate molecules on the Amazon public cloud, using a new, high-resolution algorithm Schrödinger had developed. The successful run used 51,000 Amazon cores, took about 4 hours, and cost a little more than $14,000. It would cost less than $4,000 if run today. IDC believes that at least half a dozen pharmaceutical firms are following in Schrödinger's footsteps.
- The Institute of Cancer Research (United Kingdom), the Center for Biological Sequencing (TU Denmark), and others are pursuing cures for cancer and other diseases with increasingly complex methods. Many of these, especially genome assembly and systems biology simulation, demand compute platforms beyond the standard commodity clusters.
- About 30 centers in the world are using proton radiotherapy, a very promising approach to avoiding second episodes of cancer following initial therapy. The benefit has to do with the ability to "sharp shoot" at the tumor using a Monte Carlo simulation model, followed by a whole-body Monte Carlo dose reconstruction to see how the radiation dose was distributed.

- For cost and growth reasons, insurance giant GEICO (Chevy Chase, Maryland) moved to automated price quotes on the phone. The company needed to provide quotes instantaneously, in 100 milliseconds or less. Their enterprise computers couldn't perform these calculations nearly fast enough on the fly. GEICO's solution was to install an HPC system and every weekend run updated quotes for every product, for every adult and every household in the United States. That takes 60 wall-clock hours today. The phones tap into the stored quotes and return the correct one in less than 100 milliseconds.

- The Arizona-based University of Phoenix is a predominantly online university that is owned by the Apollo Group. The university's enrollment is close to 300,000 students. To maintain and grow that enrollment, the university targets millions of prospective students using advanced mathematical models and algorithms in combination with HPC.

24.8.9 An Extreme Application: China Will Use HPC to Help Manage Its Internet of Things

At the October 2013 ISC Big Data conference in Heidelberg, Chinese Academy of Sciences (CAS) official Dr. Zhiwei Wu described a 10-year CAS research project that involves 19 Chinese institutions in planning for big data in China's mainstream markets in the era 2020–2030. A key target of the initiative is sensors in appliances and other home-based devices—an important component of the so-called Internet of Things. Today, manufacturers put one sensor in every device, and by 2030, the average Chinese home will have 40–50 devices that together generate 200 terabytes of data per year. Multiply that figure by 500 million homes and you get 100 zettabytes of new data annually.

China is already installing a smart grid and smart meters. In the future, web searches will become grid searches, such as to identify the top 100 "green" households in Beijing. The HPC architecture for managing this vast cloud-and-device network is dubbed REST 2.0. The planners foresee it including hundreds of exabyte-scale, billion-thread servers; tens of thousands of petabyte-scale servers; and millions of smaller servers. Dr. Wu said CAS is developing a new server that will scale to 1 billion threads, with a dramatically simplified architecture and hardware/software stack. CAS is also developing new storage devices that aim to improve power efficiency while meeting latency, scalability, and resiliency needs. Storage innovations are in the areas of stable sets, metadata clustering, and network RAID.

24.8.10 Vendors and the HPDA Convergence Market

The HPDA vendor scene is becoming increasingly heterogeneous and vibrant. The analytics side of the formative HPDA market is where traditional HPC users and first-time commercial adopters are converging most rapidly. Established vendors that have served each of these customer groups are exploiting this convergence by following their buyers into the new HPDA analytics territory.

Companies that once focused on commercial business analytics and business intelligence, such as LexisNexis, Oracle, SAP and SAS, have been forming high-performance analytics teams and adopting tools and methods that originated in HPC, such as parallel processing, industry-standard clusters and grids. They are doing this primarily to meet the escalating high-end analytics needs of existing customers. Contemporary HPC technologies are finally enabling companies, government agencies, and others to realize their unfulfilled 1990s dreams of turning mountains of data into actionable knowledge—often in near real time.

In the other direction, HPC stalwarts such as Cray and SGI are taking aim at both HPC and commercial markets by offering systems with special graph analytics capabilities that clusters can't match. The major hardware OEMs (HP, IBM, Dell, Bull, Fujitsu, et al.) and their storage, networking, and software counterparts have prior, separate histories in HPC and commercial markets and are learning how to fuse this experience to address the converging HPDA market efficiently.

In some cases, the convergence of HPC and commercial enterprise computing triggers schizophrenic behavior, as IT vendors and users struggle to bridge these formerly separate worlds. The applications themselves often cross this barrier along a seamless continuum, but the vendor and user organizations may not be ready to follow suit. They may not have HPC groups yet, or they may have enterprise and HPC groups that are not used to working closely together.

Smaller first-time adopters come in many shapes and sizes, including small businesses. Examples here include horizontal players such as the aforementioned cloud-based Cycle Computing (New York) and Nimbis Services (McLean, Virginia), as well as Emcien (Atlanta, GA), whose cloud-based fraud detection algorithm is used everywhere from classified government to retailers. An up-and-coming vertical player is Apixio (San Mateo, CA), which applies graph analytics to heterogeneous medical data to assess risks, help manage patient populations, and balance quality and costs for large providers.

24.8.11 HPDA Worldwide Market Size and Forecast

IDC forecasts that revenue for HPDA-focused servers will grow robustly (13.9% CAGR), increasing from $739.4 million in 2012 to approach $1.4 billion in 2017 (see Table 24.1). Because HPDA server revenue starts as such a relatively small chunk of overall HPC server revenue, the near doubling of HPDA server revenue increases its share of the overall revenue only from 6.5% in 2011 to 9.2% in 2017. HPDA storage revenue will approach $920 million in 2017.

24.8.12 A Major Challenge: Data Movement and Storage

At the system level, the HPC market is entering a kind of perfect storm. For years, HPC architectures have tilted further and further away from optimal balance between processor speed, memory access, and I/O speed. As successive generations of HPC systems have upped peak processor performance without corresponding advances in per-core memory capacity and speed, the systems have become increasingly compute centric, and the well-known "memory wall" has gotten worse. Now comes the HPDA era that will require superb memory and I/O capabilities, often with far less need for computing prowess.

HPDA workloads are emerging relatively rapidly. Partly as a result of this, storage is the fastest-growing segment in IDC's 5-year HPC market forecast. IDC predicts that the storage segment will grow at a robust 8.9% CAGR, from $3.7 billion in 2011 to $6.0 billion in 2017 (Table 24.2). That amounts to a 51% revenue jump in 5 years.

Emerging data-intensive problems are exposing more limitations of established HPC architectural designs—not just in the memory wall itself but also in the way existing, compute-centric architectures handle data movement throughout the system. It is important to make advances here,

TABLE 24.1 Worldwide HPC and HPDA Server Revenue, 2009–2017 ($M)

	2009	**2010**	**2011**	**2012**	**2013**	**2014**	**2015**	**2016**	**2017**	**2012–2017 CAGR (%)**
HPC server	8,637.1	9,504.3	10,300.0	11,032.0	11,397.4	12,371.2	13,484.6	14,620.9	15,441.0	7.0
HPDA server	535.1	602.9	672.5	739.4	785.8	881.2	1,109.0	1,253.2	1,416.0	13.9
Share of HPDA server (%)	6.2	6.3	6.5	6.7	6.9	7.1	8.2	8.6	9.2	

Source: IDC 2014

TABLE 24.2 The HPC Market Beyond the Servers: The Broader HPC Market

Worldwide HPC Server, Storage, Middleware, Application and Service Revenues ($M)				
	2011	**2012**	**2017**	**CAGR (%) (12–17)**
Server	10,300	11,098	15,441	6.8%
Storage	3,664	4,059	6,008	8.2%
Middleware	1,147	1,254	1,568	4.6%
Application	3,370	3,621	4,837	6.0%
Service	1,801	1,877	2,368	4.8%
Total	20,282	21,909	30,223	6.6%

Source: IDC 2014

or data movement for emerging HPDA problems could become frustratingly slow and expensive.

IDC research shows that the HPC community plans to deploy two main strategies for addressing the data movement challenges. The first is to accelerate data movement via more capable interconnect networks. Think here not only of interconnect vendors such as Mellanox, and Cisco but also of initiatives by processor vendors to move up a level of integration to provide capable fabrics. AMD's SeaMicro initiative comes to mind, along with Intel's acquisitions of QLogic's InfiniBand assets and Cray's proprietary interconnect assets. The second approach is to reduce data movement at all levels.

The shift toward better compute/data movement/storage balance will require a reorientation of planning and purchasing practices. Storage and data movement can no longer remain secondary considerations. The goal of using available budgets to maximize peak and LINPACK flops (macho-flops) will need to give way over time to a more singular focus on user requirements for sustained performance and time to solution. Already, one leading site, NCSA, has declined to submit HPL numbers for the Top500 rankings. The Top500 list will remain valuable for census-tracking large systems and trends affecting them over time, but the shift away from strong compute centrism will make it even more important to develop more balanced benchmarks for HPC buyers/users.

24.9 CONCLUSION

IT markets have a strong penchant for proclaiming "The Next Big Thing," but many of these developments fail to live up to their hyperbolic marketing claims. The global HPC community has learned to view "this-will-change-everything" claims with a strong dose of skepticism (along with hope that the new trends will be useful).

In this regard, HPC Big Data is fundamentally different from most new arrivals on the IT scene. For one thing, data-intensive computing has been an important part of the HPC market for decades. Existing use cases using long-standing numerical modeling and simulation methods have been growing by leaps and bounds, as evidenced by the annual doubling of storage capacities at some HPC sites, and by HPC storage revenue, which has been expanding at a 2%–3% higher CAGR than the healthy growth rate for HPC server revenue.

Hence, even if newer data-intensive methods and use cases do not proliferate as quickly or as broadly as expected, IDC believes that a strong future is assured for the HPDA market, based solely on the prospects for existing uses and methods. And newer methods and use cases are already beginning to add revenue to the HPC Big Data mix. In summary, it is certain that the HPC Big Data market will continue growing—the main question is the extent to which growth will be augmented by newer use cases, and by newer methods applied to existing and new uses.

For several reasons, a potential brake on the growth of the HPC Big Data market is data movement. First, processors used in HPC systems typically can handle far more data than the memory subsystems can feed them—this is the so-called memory wall. Second, storage access density improvements have seriously lagged advances in areal density for years; response times can be a limiting factor when many users concurrently request data. Finally, from an energy-use perspective, moving data can be considerably more expensive than performing calculations on the data. A single floating point operation might consume 1 picojoule of energy, but moving the result could cost as much as 100 picojoules.

All things considered, however, it is difficult to conclude that HPDA has anything but a bright future. Some of today's most promising HPDA use cases, such as fraud detection and personalized medicine, could be worth hundreds of millions of dollars each if and when they are broadly implemented. As HPDA evolves from today's static searches for known database entries to the emerging era of dynamic discoveries of unknown patterns, the value of HPDA—and its HPC enabling technologies—is likely to skyrocket. That should also benefit the worldwide HPC community and market.

24.10 RELATED RESEARCH

Additional research from IDC in the technical computing hardware program includes the following documents:

- The Economic Value of HPC in Science and Industry: HPC User Forum, September 2012, Dearborn, Michigan (IDC #237182, October 2012).
- Big Data in HPC: HPC User Forum, September 2012, Dearborn, Michigan (IDC #237180, October 2012).
- How Nations Are Applying High-End Petascale Supercomputers for Innovation and Economic Advancement in 2012 (IDC #236341, August 2012).
- Tokyo Institute of Technology: Global Scientific Information and Computing Center (IDC #236243, August 2012).
- Shanghai Supercomputer Center (IDC #236245, August 2012).
- Petascale Supercomputing at the University of Tokyo (IDC #236292, August 2012).
- HPC Application Leadership at the Supercomputing Center of Chinese Academy of Sciences (IDC #236281, August 2012).
- India Broadening Access to Supercomputing (IDC #lcUS23621912, July 2012).
- Institute of Process Engineering, Chinese Academy of Sciences (IDC #236204, July 2012).
- How Fujitsu Built the World's Fastest Supercomputer in Record Time and Ahead of Schedule (IDC #235733, July 2012).
- Europe Sharpens Its Focus on Exascale Computing (IDC #lcUS23555112, June 2012).
- IBM Returns Supercomputer Crown to United States (IDC #lcUS23547812, June 2012).
- Potential Disruptive Technologies: HPC User Forum, April 2012, Richmond, Virginia (IDC #234742, May 2012).

- The Broader HPC Market: Servers, Storage, Software, Middleware, and Services (IDC #234682, May 2012).
- HPC End-User Site Update: RIKEN Advanced Institute for Computational Science (IDC #233690, March 2012).
- National Supercomputing Center in Tianjin (IDC #233971, March 2012).
- Worldwide Data Intensive–Focused HPC Server Systems 2011–2015 Forecast (IDC #232572, February 2012).
- Exploring the Big Data Market for High-Performance Computing (IDC #231572, December 2011).
- High-Performance Computing Center Stuttgart (IDC #230329, October 2011).
- Oak Ridge $97 Million HPC Deal Confirms New Paradigm for the Exascale Era (IDC #lcUS23085111, October 2011).
- China's Third Petascale Supercomputer Uses Homegrown Processors (IDC #lcUS23116111, October 2011).

CHAPTER 25

Conclusions

Synergy between Science and Industry with High-Performance Computing

Anwar Osseyran and Merle Giles

CONTENTS

25.1 INTRODUCTION

We have shown in this book how in the past decades, high-performance computing (HPC) has widely proven to play a crucial role in technology innovation and scientific discovery in almost all areas of science and industry. Although national defense programs were the initial drivers of investments in supercomputing, science and engineering increasingly adopted computational methods to support discovery and technological innovation.

25.2 FROM DEFENSE TO OPEN SCIENCE

The rapid development of semiconductor technology led to an exponential increase in computational capabilities and offered scientists and engineers an alternative to experimentation that was capable of complex exploration outside the laboratory. In the early 1980s, only a quarter of the world's HPC capacity was used for military purposes. Remaining uses included open research, physics and engineering, oil and gas exploration, and nuclear research. Investments in supercomputing technologies quickly became associated with scientific and technological competitiveness, and governments and universities became the biggest customers of the supercomputing suppliers by the end of last century. Nationally funded supercomputing centers chose to associate with universities, which played a pivotal role in advancing science and supporting industry in technological innovation. Scalable HPC applications (such as the NASA's NASTRAN) were used by industry researchers across the globe.

25.3 FROM OPEN SCIENCE TO VIRTUAL PRODUCT DEVELOPMENT

In more recent times, the global financial crisis made it clear to governments and academia that funding for e-infrastructure and services can only be sustained if there is an economic return on these huge national investments. In this book, we show that HPC has fostered significant advances in science and engineering and that the deployment and use of advanced modeling and simulation through HPC is central to improved industrial products and services. Partnerships between science and industry will become even more valuable in the modern digital age, during which HPC capabilities will make it possible to conceive, develop, and test entire products and assemblies in a virtual world. Petascale and exascale simulation capabilities, combined with Big Data analytics will make products more competitive in silico before physical prototyping and product deployment.

25.4 INDUSTRY NEEDS SCIENCE AND VICE VERSA

Advancements in multidisciplinary science and digital technologies benefit society directly by helping industry decrease time-to-market, product development costs, and material reuse. Simultaneously, industry benefits multiple domains of science by demanding increased realism and capability and by investing in multidisciplinary science, thus stimulating technology innovation. Synergy between science and industry is therefore valuable and even irrevocable as scientists need to finance the growing costs of their research and infrastructures and get inspired about which challenges they should tackle, and as industry needs to deploy cutting-edge technologies to remain competitive. Most ivory towers of isolated scientists have collapsed as have most of the industries that failed to deploy scientific advances in their own products and services. Partnerships between science and industry has proven to lead to mutual benefits, accelerating exchange of ideas and information, expanding markets, improving design, comfort and sustainability of products, and increasing societal benefits of science and technology.

25.5 FACTORIES OF THE FUTURE WITH HPC

Manufacturing is making its comeback as a high-tech sector addressing the challenge of efficiently producing more customized products in a prosumer* (producer–consumer) world. Factories of the future are expected to be flexible, produce more goods with less energy and material, and be accessible to the consumer. Innovation in manufacturing requires the deployment of scientific simulation and modeling techniques of products but also manufacturing processes, integrating HPC in the workflow and using HPC during the whole life cycle of products. Big Data also helps to anticipate customer needs and behaviors enabling factories to produce the right products just-in-time, minimizing waste and optimizing the use of material and energy. This has led to various public–private initiatives in the United States, Europe, and southeast Asia. These initiatives are focused on bringing together HPC infrastructure, scientific and technological know-how, HPC applications, and industrial and market expertise while driving increased use of multidisciplinary and multiscale predictive modeling.

Small- and medium-sized enterprises (SMEs) that represent the majority of the manufacturers in the world do not possess the R&D depth and financial wherewithal to invest in the leading edge of these developments and must therefore be integrated into an advanced computing ecosystem to avoid losing out on modern digital innovations. Large manufacturers are increasingly aware that they are part of an ecosystem and that they must pull their, mostly SME, suppliers with them. The above-mentioned public–private programs would enable SMEs to advance more rapidly by reducing entry barriers and providing easy and affordable access to HPC infrastructure, expertise, applications, and services.

* http://en.wikipedia.org/wiki/Prosumer.

Cloud computing and data mobility will play crucial roles in offering advanced digital modeling, simulation, and analysis on-demand, as it becomes even more ubiquitous. Some would say that "HPC in the cloud" offers the manufacturing masses advanced capabilities that used to be the exclusive domain of HPC specialists, which can be expected to increase innovation and more capably coordinate with science, engineering, and product assembly at large companies.

25.6 USING THE POWER OF BIG DATA

Big Data is igniting new scientific and industrial breakthroughs by deploying HPC and data analysis capabilities. It is now possible to generate, collect, store, and process massive amounts of structured and unstructured data to reveal intrinsic near-real-time value to science, business, and society. Google, Facebook, Yahoo!, Twitter, and Microsoft paved the way for others to collect satellite information, maps, tweets, images, videos, and social interactions, developing new innovative services and making a deep impact on society, business, and even science. The so-called Fourth Paradigm of Science is based on the full exploitation of those massive volumes of data, resulting in new discoveries that are not only based on theory, experiments, or modeling. High-value data correlations and predictions will pave their way into science alongside causation through theory, experimentation, and modeling.

Commoditization of HPC infrastructure in conjunction with open-source software and platform interoperability makes it possible to deploy data analytics to cope with data variety, volume, and velocity on top of large-scale simulation and modeling and to provide the insight hidden in the growing data lakes around us. HPC techniques like virtualization and cloud computing made it possible to scale capacity-on-demand and dramatically reduce costs of data storage and processing. Virtualization techniques abstract the underlying physical hardware, offering higher-level services such as cloning a data node, high availability to a specific node, and user-controlled provisioning. Cloud computing complements these techniques by offering additional HPC compute and data resources on demand.

The potential applications of Big Data are in almost every sector, including science, health, e-commerce, government, energy, environment, finance, and manufacturing. The ultimate challenge, however, does not seem to be the technology itself, but how to develop enough skills to make effective use of the new technologies to extract value out of huge amounts of data. Promises of Big Data include innovation, growth, and long-term sustainability while threats include breach of privacy, property rights, data integrity, and personal freedom. Having science and industry working closely together will accelerate the exploitation of data in an open and transparent manner, delivering the true promise of Big Data.

25.7 HPC USE CASE STUDIES

Various case studies of industrial use of HPC have been presented in this book. These case studies focus not only on the technological advances offered by HPC but also on how HPC helps lower design and production costs while maximizing the acceptability of the product in the marketplace. The presented case studies cover various industrial areas and deal with multidisciplinary scientific domains, constituting a worthy challenge not only to engineers and their companies but also to the researchers and scientists involved.

The business rationale for the use of HPC has shown that despite the relative high costs of HPC infrastructure and expertise, optimization of high-tech products can lead to significant cost savings of time, energy, and material. For instance, a small percentage improvement in the fuel efficiency of GE's low-pressure turbines or the Rolls-Royce three-shaft gas turbines would lead to hundreds of millions of dollars in annual fuel savings across the aviation industry. The potential savings are possibly an order of magnitude greater if similar advances are applied to other products that are directly or indirectly dependent on energy.

The business case for HPC in the oil and gas industry has always been a "no-brainer." Seismic imaging was, as presented in the use cases of Repsol and BP, the most dominant application of HPC in the 1980s. HPC-enabled 3D imaging drove immense improvements in exploration efficacy of oil and gas companies, improving the production replacement ratio (of discovered to produced oil) by a factor of three, the costs of finding oil by a factor of eight, the new resources discovery by a factor of five, and the exploration drilling success rate by a factor of three. Oil and gas companies recognize that HPC is a critical instrument for lowering risk and uncertainty.

The use case of HPC in the automotive industry was well presented by Porsche and Renault in this book. The automotive sector was also one of the first industrial sectors to adopt HPC. The technological impacts of HPC include improving passive and active safety through better and realistic crash simulations; optimizing weight, aerodynamics, acoustics, durability and combustion yield; improved fuel consumption and use of sustainable energy; and enabling efficient digital design, manufacturing, and shorter time-to-market. HPC has become increasingly embedded in car development processes to comply with ever-demanding requirements regarding safety, construction, and emissions and coping with dynamic economic, market, and competitive factors. The technical disciplines cover structural mechanics, material science, fluid dynamics as well as robustness, numerical optimization, and multidisciplinary simulations.

25.8 APPLICATION SOFTWARE IS KEY IN HPC

As we can conclude from the various use cases presented in the book, application software is a key component in utilizing large-scale supercomputing and needs to be considered equal to the challenges of hardware, data storage, and networking. Several case studies shared here are prime examples of how teams of developers and users look beyond processors, FLOPS, and interconnects to compilers, code architecture, and algorithms to extract the most performance from these massive machines. The implementation of Moore's law on hardware performance has taken a different approach, now focusing on more processor cores on silicon chips rather than speeding up individual hardware components. This proliferation of more cores on a single processor chip has ushered in an urgent need to program applications in parallel.

Increased attention, therefore, is needed in software engineering and programming, complementing the classical aspects of computer science and computational methods. University HPC centers should take the lead in these efforts in the absence of national and continental policies. This leadership should include the establishment of high-value career paths for software engineers and programmers, separately from the pure disciplines of science and engineering. The scarcity of parallel programming skills in the context of domain science and engineering is stifling growth in innovation through digital techniques.

Another point of attention is the need for open collaboration between universities and the industry both on structural programming allowing scientists to learn how to develop sustainable codes that survive the completion of the PhD of the researcher and also enable the software vendors to restructure their legacy software to cope with new algorithms and petascale architectures. For instance, scientific and technical software lifecycles are more than an order of magnitude longer than the software lifecycles in video gaming, and national investments in hardware have dwarfed investments in software. The result is that scientific and industrial communities are struggling with legacy applications that do not scale in modern multicore and many-core systems. A radical reinvestment of legacy software is needed, and the authors suggest that this investment begins with widely accepted open-source and commercial applications. This will require collaboration between scientific and industrial communities and will benefit broad bases of user communities in a wide variety of countries. The application examples shared in this book that show breakthrough performance will hopefully increase attention and investment.

25.9 HPC ACROSS THE GLOBE

HPC has been recognized in most industrialized countries as an instrument of change for scientific and engineering communities. Comprehensive and pragmatic national HPC strategies have been developed and deployed in the United States, Europe, China, Japan, South Korea, India, and Russia. The United States historically led the world with the strength of its hardware vendors, yet Japan and China in recent years have secured number one positions on the TOP500® supercomputer list.

With a balanced approach, U.S. HPC strategy supports HPC suppliers, service providers, and users through its DARPA program and large federal agencies in defense, energy, and science. In Europe, almost all European countries are partners in the PRACE project, focused on developing a pan-European HPC-provisioning pyramid with four countries hosting leadership machines (Germany, France, Italy, and Spain) and others hosting Tier-1 national HPC centers. Germany is leading with various Tier-0 machines that service users across the European continent. France and the United Kingdom are also focusing their national efforts and investing heavily in the deployment of HPC in industry and manufacturing. Switzerland has at the moment the largest European supercomputer dedicated to scientific research. In Asia, China has taken in 2013 the lead from Japan with number one position on the TOP500® list and the second-largest national HPC investment following the United States. Other countries like South Korea, India, and Russia are investing heavily in HPC to support their own national scientific and industrial positions.

A growing difference is appearing between East and West attitudes toward industry, with the gap between science and industry appearing much smaller in the East. India, China, South Korea, Japan, and Russia seem to have better coordinated HPC programs for national industry, whereas a revitalization of manufacturing in Europe and the United States has failed so far to incite industrial policies and investment in the West. In all corners of the world, investment is absent in resources and talent targeted at improving human health and stemming disease. Dr. Klaus Schulten's success on Blue Waters[†] should inspire us all to pay more attention to this sector.

Despite the predominance of national and continental funding of HPC, most supercomputers remain open to researchers from abroad as most countries are aware of the importance of global collaboration and of the magnetic effect of HPC infrastructure on talent and technological innovation. Academic scientific communities have organized themselves across national boundaries and are seeking increased deployment of HPC resources in their own domains.

25.10 TOWARD THE BLUE OCEAN WITH HPC

Publishers Weekly, in a review of the *Blue Ocean Strategy* book by France's INSEAD business school professors W. Chan Kim and Renée Mauborgne, describes the blue ocean metaphor elegantly as the kind of expanding, competitor-free markets that innovative organizations can navigate. Unlike "red oceans," which are crowded with competitors, "blue oceans" represent an opportunity for highly profitable growth. However, the dominant focus of strategy over the past 25 years has been on competition, finding new ways to cut costs and grow revenue by taking away market share from others. A "blue ocean" strategy aims to resolve this issue by stimulating organizations toward value innovation and to create and capture new demand, focusing on the big picture rather than financial numbers alone. In other words, the INSEAD authors argue that lasting success comes from creating blue oceans and new growth.

To be sure, the world of supercomputing and HPC in science and engineering is competitive. In May 2014, Cisco's CEO John Chambers stated that "brutal, brutal" consolidation is coming in the information technology sector. Coincidentally, an example of this began in a highly visible manner within the HPC community with France's Atos bid to purchase Groupe Bull. Next-generation supercomputers will require huge investments in infrastructure, software, and expertise—perhaps in ways that no single country or nation can afford alone. France's potential merger of a cloud and HPC provider would certainly qualify as a merger to watch.

The business potential of some of today's most promising use cases, such as smart energy, manufacturing, and medicine, is enormous. But while returns on investment are likely to be high, attention needs to be paid to reducing risks. Numerous user communities are now global, and

[†] http://www.ncsa.illinois.edu/news/stories/BWfriendlyuser/.

indeed, academic science communities around the world regularly collaborate on research. Collaboration is not observed, however, in the digital infrastructure, data resources, and centralization of talent. If user communities adopt a collaborative blue ocean approach, the transformative power of HPC will boost global macroeconomic growth and innovation, skipping the bloody red oceans of rivals.

As authors and coeditors, we wish to share with you our extremely optimistic view of the blue ocean of supercomputing and extreme data. In contrast to the red ocean of vendor consolidation, it is our opinion that there is perhaps nothing more powerful than the promise of discovery and innovation through the use of very advanced, large-scale HPC. If, indeed, the wealth of nations is tied to industrial wealth, as suggested in the Executive Summary, then supercomputing centers around the globe are well positioned to lead the way to the blue ocean. Central to existing collaboration are domain scientists. Secondary, but no less important, is the impact that HPC centers have on highly scientific and technical industries, many of which reach far beyond their provincial borders to conduct business. Importantly, the world's top companies know no borders but for the legal and tax implications of multinational production and distribution.

It is time for the world's leading companies and the world's leading technical centers of discovery and innovation to work together on the most intractable challenges to society. By so doing, we can expect the following:

- To collaborate in ways that reduce the singular risk of doing it alone
- To collaborate on global resource provisioning that will embrace precompetitive invention and innovation
- To increase commercial access to powerful resources that cannot be replicated elsewhere
- To optimize scarce resources and lead nations and governments to muster scientific and engineering resources in the blue ocean
- To dominate the forces of human health, energy, finance, distribution, and manufacturing in ways that create and capture growth

In the spirit of American baseball, as we swing for the fences, we will occasionally hit home runs. If we do not swing for the fences, we will be forever remembered for our batting average, which is, after all, just average.

Index

D

E

F

G

H

J

K

L

M

N

O

R

S

T

U

V

W

X